"十三五"普通高等教育本科部委级规划教材

服装设计英语

董晓文 位 巍 言 婷 陈素英 编著

中国纺织出版社有限公司

内 容 提 要

本书顺应高等院校对人才培养的趋势与需要，为服装与服饰设计专业提供内容全面的专业英语学习参考，且图文结合、双语对照。本书内容涵盖中西服装演变史，服装设计的款式、色彩、面料三要素，服装配饰，女装、男装与童装设计，服装效果图，服装系列研发与生产过程等。同时配有标准美式英语的外教朗读音频，辅助学生学习与发音。本书可作服装与服饰设计本科专业的教学与学习之用，以及服装设计专业人士或爱好者的阅读之用。

图书在版编目（CIP）数据

服装设计英语 / 董晓文等编著 . -- 北京：中国纺织出版社有限公司，2020.11

"十三五"普通高等教育本科部委级规划教材

ISBN 978-7-5180-7550-8

Ⅰ . ①服… Ⅱ . ①董… Ⅲ . ①服装设计—英语—高等学校—教材 Ⅳ . ① TS941.2

中国版本图书馆 CIP 数据核字（2020）第 112129 号

策划编辑：苗 苗　　责任编辑：金 昊
责任校对：楼旭红　　责任印制：王艳丽

中国纺织出版社有限公司出版发行
地址：北京市朝阳区百子湾东里 A407 号楼　邮政编码：100124
销售电话：010 — 67004422　传真：010 — 87155801
http://www.c-textilep.com
中国纺织出版社天猫旗舰店
官方微博 http://weibo.com/2119887771
北京通天印刷有限责任公司印刷　各地新华书店经销
2020 年 11 月第 1 版第 1 次印刷
开本：787×1092　1/16　印张：13.75
字数：237 千字　定价：48.00 元

前 言
Preface

　　随着世界经济和服装市场的全球化，服装业迎来了前所未有的机遇和挑战。在这种趋势下，培养既精通专业又能用外语交流的复合型人才成为行业的需要。而且目前高校各专业对专业外语或双语教学也越来越重视，并增加了这类课程在培养方案中的比重。服装与服饰设计专业也顺应这一趋势，逐渐重视学生的双语交流能力，意在将他们培养成高层次人才，而培养人才就需要一本适合的教材。目前国内的同类教材内容多集中在服装工程制板和外贸函电等方面，以《服装设计英语》命名的教材非常少，术业有专攻，教材内容的选择往往集中于作者擅长的领域。从现有情况来看，《服装设计英语》教材的编写与研究还存在一定的空白与需求。因此编者便以此为契机，旨在编写一本更贴近服装与服饰设计专业的特点、更适合本专业学生学习和阅读的教材。

　　教材编写从本专业的特点出发，内容全面、难易适中。其包含了中西服装演变史、服装设计要素（款式、色彩、面料），服装配饰、服装分类设计、服装效果图、服装系列研发与生产过程等内容，共分为十四章。每一章节都进行了比较深入的讲解，侧重于学生的基础学习。让学生了解服装设计的基础英语内容与特定词汇，为将来进一步的学习与应用奠定基础。课后练习多为课堂讨论内容，加强学生之间的交流与互动，让学生学以致用，让教师寓教于乐。

　　第二章"中国服装演变史"作为专业英语内容首次纳入编写范围。中国服饰文化作为中国传统文化灿烂的一页也是对外文化交流的一部分，也需要学生学习和了解，且中国服装史一直是设计专业学生的必修课程，所以编者特地增加了这一部分。本教材按照中国的朝代顺序，对每一朝代的代表性服装配饰做了讲解。为便于学生接受，在语言编写上尽量通俗易懂，并配有相关的图片参照。此外，本教材也首次对电脑绘画方面的内容进行了编写。在国内同类教材中，服装CAD（服装辅助设计）讲解的较多，而服装电脑绘画则几乎没有涉及。教材对电脑图像的分类、绘图软件的特点介绍，以及绘画方法等进行了阐述，让学生能够对其绘图过程做一个基本的英文表述，以便于日后更好地交流。

　　教材采用双语对照的模式编写，以便更好地帮助学生理解和教师教学。为保证英文语言的规范性，编者阅读了大量原版教材，择其适用部分编入章节，并请专业外语老师做语法审核，以保证教材内容的准确性。

教材中绘有200多张服装与配饰的款式图，多数为矢量绘图，绘制清晰明确。因为服装设计专业的单词术语多为具体实物，又根据艺术设计专业的学生形象思维能力强、思维活跃的特点，采用图文结合的方式，尽量将抽象的文字形象化，让学生对照图片学习课文，增强学习的直观性和趣味性。

为帮助和纠正学生发音，同时便于教师教学，编者特地请美国外教与专业的录音棚为全篇英文制作了标准的美音音频，将教材做成有声版，学生可以借助音频练习口语和听力。

本教材由青岛大学纺织服装学院教师董晓文主编，青岛大学外语学院教师位巍、山东轻工职业学院教师现东华大学在读博士言婷、青岛大学纺织服装学院副教授陈素英副主编。其中第一至第八章，第十一和第十二章由董晓文编写，并绘制文中插图（署名除外）；第九章和第十三章由董晓文和陈素英编写；第十章和第十四章由言婷和山东轻工职业学院教师纪静编写，并绘制插图；位巍做全篇英文部分的润色和校正；最后由董晓文统稿。青岛大学梁晓琪和陈明慧同学为第三章绘画插图，并为第十一章和第十三章拍摄照片。青岛大学邓跃青、韩蔚老师亦为教材提供图片。编者在此对以上老师和同学表示最真挚的感谢，同时，也要感谢网络图片资源的作者和提供者。也许教材的编写因编者水平和时间有限而有不准确之处，还请读者不吝赐教，在此表示感谢。

编者

2020年1月

目 录
Contents

The Introduction to Fashion Design
服装设计概述

课题名称： The Introduction to Fashion Design

课题内容： 服装设计概述

课题时间： 2课时

教学目的： 让学生学习并了解服装、服装设计的概念，影响服装设计的因素，以及如何寻找服装设计灵感等内容。

教学方式： 结合PPT与音频等多媒体教学课件，以教师课堂讲解为主，学生随堂练习与小组讨论为辅。

教学要求： 1. 掌握相关词汇。

2. 了解不同服装词汇的含义和区别，用英语描述服装设计的含义、影响因素和寻找灵感的方法。

课前（后）准备： 课前预习关于服装和服装设计概念的内容，浏览课文；课后记忆单词，查阅相关资料，并口述文中内容。

1.1 The introduction to clothing 服装的概述

1.1.1　The function of clothing 服装的作用

Clothing refers to those articles worn on the human body. It can protect the human body; keep it warm; make people comfortable, decent and pleasant. It functions as a silent communication system that provides basic information about age, gender, marital status, occupation, religious affiliation, and ethnic background for everyday or special occasions. What a person wears can also indicate personality characteristics and aesthetic preferences.

服装是指穿在人身上的物品。它可以保护身体，保暖，让人感到舒适、体面和愉悦。它是无声的交流系统，在日常或特定场合中可以传达出人们的年龄、性别、婚姻状况、职业、宗教信仰和民族背景等基本信息。一个人穿什么还可以显示出他的性格特征和审美偏好。

1.1.2　Some expressions relating to clothing in English 英语中"服装"的相关用语

Fashion 时装/时尚

Fashion refers to a popular style of clothing at a particular time or place. Fashion is not only defined in the contemporary era, but there was specific fashion produced in every historical era. The newer styles comparing with those common and stable ones in a certain period are all part of fashion. Fashion is characterized by change; it encompasses all forms of self-fashioning, including street styles, as well as so-called high fashion created by fashion designers. At present, there are more and more people following fashion, mainly due to the development of society and economy, the change of ideas, and the popularity of the internet.

时装是指在一定时空里流行的服装样式。时装不仅限于当代，每个历史时期都有特定的时装。在某个时期，相对于那些常规而稳定的服装而言，较为新颖的款式都属于时装。时装的特点是变化，包括各种形式的自我塑造，既包括由时装设计师创造的高级时装，也包含街头时尚。现在，追求时尚的人越来越多，主要是因为社会和经济的发展，人们观念的变化，以及互联网的普及。

Costume 服装/装束/戏服

Costume describes garments of many types, particularly when worn as an ensemble. It refers to the clothing items, accessories, and makeup, for actors, dancers, and people dressing up for special events such as Halloween, masquerade balls, or Carnival; it can also refer to ensembles of clothing worn by minorities.

"costume"可以形容多种类型的服装，尤其指整套的装束。它涉及服装、饰物和化妆，是指演员、舞蹈演员和人们在特殊活动如万圣节、化装舞会或狂欢节中所做的装扮；还可以指少数民族的装束。

Dress 服装/装扮

Dress indicates a woman's one-piece garment, or a category of garments such as "holiday dress" or "military dress"; or as a general reference to an individual's overall appearance and various identities. Dress indicates the process of covering, adorning, and modifying the body. The act of dress involves all five senses and encompasses more than wearing clothes. It includes arranging hair, applying scent, and cosmetics, as well as putting on clothing and jewelry.

"dress"指女性连身的衣服，或一种服装类型，如"节日服装"或"军礼服"；或作为个人整体形象和不同身份的一般参考。装扮指穿着、装饰和修饰身体的过程。装扮的动作不仅是穿着，还会涉及五感，除了穿衣服、戴首饰以外，还包括打理头发、喷洒香水和化妆。

Further reading: other glossaries of clothing

延伸阅读：其他服装词汇

clothes	[pl.] the things that people wear 衣服
clothing	[U] clothes, especially a particular type of clothes; "clothing" is more formal than "clothes" 衣服，尤指某种类型的服装；clothing 比 clothes 更正式
wear	[U] (usually in compounds) clothes for a particular purpose or occasion, especially when it is being sold in shops/stores（通常构成复合词）指某种用途或场合穿的衣服，尤指商店售卖的衣服
apparel	[U] the clothes sold in shops or stores; (formal) formal clothes worn on a formal occasion 商店出售的衣服、成衣；正式场合穿的服装
attire	[U] (formal) clothes 服装、盛装
garment	[C] (formal) a piece of clothing; only used in formal or literary contexts 一件衣服；只用于正式或文学语境
garb	[U] the clothes worn by a particular type of person; unusual clothes 某类人穿的特定服装；奇装异服

ensemble	[C] (usually sing.) a set of clothes that are worn together 全套服装
outfit	[C] a set of clothes that are worn together, especially for a particular occasion or purpose 全套服装、装束，尤指为某种场合或目的
gear	[U] (informal) clothes 衣服

pl. 复数名词；sing. 单数名词；U 不可数名词；C 可数名词；formal 正式用语；informal 非正式用语

1.2 The brief introduction to fashion design 服装设计简述

The purpose of fashion design is to be worn; thus the human body and the clothing are the two most essential elements of fashion design (Figure 1-1). The body provides the frame, the motion, and the inspiration for fashion design. Fashion is made by forming a 2D textile into a 3D shape; it has a special style, silhouette, size, color, and texture.

服装设计的目的是穿着，因此人体和服装是服装设计的两大基本要素（图1-1）。人

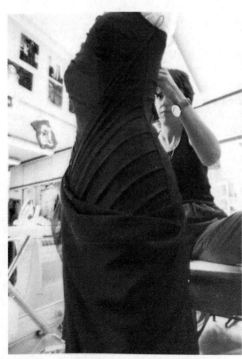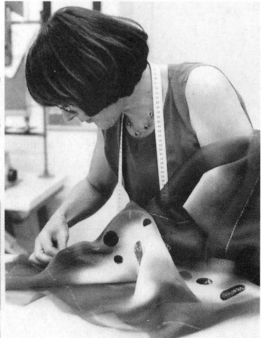

Figure 1-1　Fashion design
图1-1　服装设计

体给服装设计提供了框架、动态和灵感。服装是把二维的纺织品做成三维的造型，具有一定的风格、廓型、尺寸、色彩和质感。

Why study fashion design? The most common answer to this question is the desire to create clothing with beauty, pleasure, and meaning. Helping people feel good through the clothing they wear is an amazing experience. The meaning of fashion design of a new dress for a special event, a suit that helps someone feel confident during a job interview, or a uniform that protects an officer from danger goes beyond the stereotype of runway fashion. Fashion design is a dynamic, exciting field for those who make it their career and those who enjoy the final product. It is also a field that involves both creative and technical skills. Fashion design reflects current society as it is evolving and changing and the lifestyle preference of consumers.

为什么学习服装设计？这个问题最基本的答案是设计师渴望创造漂亮、愉悦并富有内涵的服装。帮助人们通过穿衣感到舒适，是种很棒的体验。为一场活动设计礼服，为帮助面试者提高自信设计套装，或为保护警官避开危险设计制服，这些服装设计相对T台上的时装来说更具有现实意义。服装设计对于那些把它当作职业和享受最终产品的人来说，是一个充满活力与激情的行业，也是一个兼具创造性和技术性的行业。服装设计反映出当前社会的发展与变化，以及消费者喜爱的生活方式。

What is the difference between design and art? Art mostly expresses the creator's emotion and thoughts; design, on the other hand, must respond to and meet the demands of the consumers. Fashion can rise to the level of art in the hands of a designer who has good vision and innovative ideas. However, the most important thing for designers is to understand their target consumer's taste and design products that the consumers will be fond of.

设计和艺术的区别是什么？艺术主要表达了创作者的情感和思想；设计，从另一方面讲，必须要迎合和满足消费者的需求。通过有眼光和创意想法的设计师之手，服装可以达到艺术的水准。但是，对设计师来说最重要的事情是要了解目标消费者的品位，设计出他们喜欢的产品。

Fashion design is a comprehensive practical art. It is a process that materializes creative ideas of designers including aesthetic principles, technology and production. Fashion design reflects a designer's personality, thoughts and emotion, along with the product brand concept. There is more to fashion design than the aesthetic component; it involves understanding the materials, the production, and the function of a garment. The care requirements of a garment and how it can be disposed of or recycled with minimal impact also need to be considered at the beginning of the design process.

服装设计是一门综合性的实用艺术。它是一个把设计师的创作思维物化出来的过程，其中包括美学法则、技术工艺和生产制作。服装设计反映了设计师的个性、思想和情感，以及产品的品牌理念。服装设计不仅仅是审美活动的一部分；它还包括对材料、生产和服装功能的了解。对服装保养的要求，以及如何对其进行最小影响程度的处理和回收，也需要在设计之初考虑。

Fashion designers have the power to build and create products, channel their imagination, solve problems and have fun. Designers are agents of change, because, in response to their observations, they imagine things that don't exist yet. They have a great desire to develop new design ideas that address the consumer's aesthetic, physical, and emotional needs while supporting business and social development requirements.

服装设计师有能力打造和创作产品，发挥他们的想象力，解决问题，并以此为乐。设计师是变革的倡导者，因为他们会根据自己的观察结果来设想一些尚不存在的事物。他们有着开发新设计理念的强烈愿望，以满足消费者在审美、生理和情感上的需求，同时支持商业和社会发展的需要。

There is a rush of opportunities and challenges in the field of fashion design, and the need for strong designers is critical. Designers do not sit at a desk and design pretty dresses, but need to acquaint themselves with the market situation, develop a design theme corresponding to product positioning, and search for fabrics. When developing a line, designers need to think about who they are designing for, what type of garments they are developing and for what season. Experts in areas ranging from business to textiles and information technology to psychology can contribute valuable ideas.

服装设计是一个充满机遇与挑战的行业，迫切需要优秀的设计师。设计师不是闭门造车，只管设计漂亮的衣服，而是要了解市场行情，研发符合产品定位的设计主题，并寻找面料。在研发一个产品系列的时候，设计师需要考虑给谁设计，设计什么类型和哪个季节的服装。其他领域的专家，从商业到纺织业、从通信技术到心理学都能为服装设计提供有价值的想法。

1.3 Influences on fashion design
服装设计的影响因素

1.3.1 Historical fashion 服装历史

Period silhouettes, details, fabrics, patterns, finishing techniques, and even the cultural attitudes associated with a particular era can all be materials from which a new collection is conceived. Historical fashion has a very important reference value for the current fashion design (Figure 1-2), especially fashion after the 1900s. Designers can predict trends of the next season by studying and analyzing fashion that has appeared in the past.

每一个时期的廓型、细节、面料、图案、后整理工艺，甚至与特定时代相关的文化态度都可以成为新服装系列的设计素材。服装历史对当前的服装设计有着非常重要的参考价值（图1-2），特别是20世纪以后的服装。设计师可以通过对以往服装的研究、分析，来预测下一季的流行趋势。

1.3.2 Architecture 建筑

Sometimes an architecture style may influence fashion, like Baroque and Rococo architecture. Architecture and clothing are both forms of spatial art; many design methods of architectural appearance can be translated into fashion design, both formally and conceptually (Figure 1-3).

有时建筑风格会影响服装，像巴洛克和洛可可建筑。建筑和服装都属于空间艺术；许多建筑的外观设计都可以应用到服装设计中，无论形式上的，还是概念上的（图1-3）。

1.3.3 Ethnic costumes 民族服饰

Because different regional ethnic costumes have their respective histories and cultures, there are often special meanings behind the wear, such as political, religious, and societal identifications (Figure 1-4).

Figure 1-2   A dress with some Tang dynasty's clothing elements ("Chuhetingxiang" Spring / Summer 2016 Fashion Show)
图1-2   带有唐代服饰元素的裙装（2016春夏"楚和听香"时装发布）

Figure 1-3   The appearance of the architecture is adopted on the costume
图1-3   服装上采用了建筑的外观

Ethnic costumes from around the world, such as samurai armor, traditional Tibetan textiles, and the body adornment of indigenous Amazonian tribes, not only have distinctive appearance but also particular meaning. Designers ought to know them well first before using them in their work.

由于不同地区的民族服饰有着各自的历史和文化，在服装穿着背后往往具有特殊的意义，如政治、宗教和社会的认同（图1-4）。来自世界各地的民族服饰，如日本武士的盔甲，藏族传统的纺织品，以及亚马孙土著部落的身体装饰，不仅拥有独特的外观而且具有特定的含义。设计师应先很好地了解它们。

Figure 1-4　Bohemia style dress
图1-4　波西米亚风格服装

1.3.4　Street styles 街头风格

Street styles came into prominence during the 1960s when designers wanted to express the deeper political mood of the time and the social changes that were occurring (Figure 1-5). In an effort to attract younger generations, designers such as Yves Saint Laurent launched more affordable ready-to-wear collections. Today designers often reference street styles so that they may employ ideas or images that are connected to their youth-obsessed consumer. Examples include Vivien Westwood's interpretation of Punk, and New Romantic Movements of the 1970s-1980s.

街头时尚是从20世纪60年代开始产生显著影响的，当时的设计师想要表达更深层次的政治情绪和正在发生的社会变革（图1-5）。为了吸引年轻的一代，伊夫·圣·洛朗等设计师推出了更经济实惠的成衣系列。如今的设计师经常借鉴街头时尚，以便采用与他们崇尚年轻的消费者相关的设计理念和形象。此例包括薇薇恩·韦斯特伍德对朋克风格的诠释，以及20世纪70年代到80年代的新浪漫主义运动。

Figure 1-5　Street style dress
图1-5　街头风格服装

1.3.5 Technology 科技

This is an area that is gradually proving its application to fashion design in limitless ways. Technology has made designers rethink the function of a garment, how production methods can innovate design, and how a seemingly unrelated technical area can influence fashion aesthetically (Figure 1-6). Technology's relationship with fashion design contains a breadth of application, including smart textiles that may someday adjust to the weather, garments that contain wearable technology, and new methods of production that can create design details—such as laser cutting.

Figure 1-6　The clothing is made with LED technology
图1-6　用LED技术制作的服装（Olga Noronha作品）

这个领域正逐渐证明其在服装设计上具有无限潜能。科技让设计师重新考虑服装的功能，考虑如何通过生产方式来革新设计，以及如何让看起来毫不相干的技术领域从审美上影响服装（图1-6）。科技和服装设计的关系涉及广泛的应用范围，包括某一天根据天气进行自我调节的智能纺织品，含有可穿戴技术的服装，以及可以创造设计细节的新的生产方式——如激光裁剪等。

1.3.6 Social elements 社会要素

Because humans live in society, social politics, economy, cultures, ethics and religions will influence fashion design. Whether it is Chinese traditional clothing or a Western historical costume both are branded by their contemporary social evolution. Certain social settings also deeply influence designer's awareness, consumer's lifestyle, and fashion trends.

人生活在社会中，社会的政治、经济、文化、伦理和宗教将会对服装设计产生影响。不管是中国传统服装，还是西方历史服装，都带有同时期社会的烙印。一定的社会背景也会对设计师的观念意识，消费者的生活方式和时尚潮流的趋势产生深刻的影响。

1.4 Research and inspiration
调研和设计灵感

No good design can occur without some form of research taking place. In terms of design, "research" means to do some creative investigation of visual or literal references that will inspire you. Designers need to be continually seeking new inspiration in order to keep their work fresh and contemporary, and above all, to keep them excited. This is why they are constantly observing their surroundings, even when they are not working. Designers need all kinds of information which should be gathered and interpreted quickly to be useful to them.

没有一定的调研就不会有好的设计。就设计而言，"调研"意味着对能启发你的视觉或文字资料进行有创造性的调查研究。设计师要不断地寻求新的灵感来保持设计作品的新鲜感和时代感，最重要的是，让自己永葆激情。这就是为什么他们即使在不工作的时候，也经常观察周围事物的原因。设计师需要各种各样的信息，这些信息应能快速地被收集和整理成对他们有用的东西。

Where are the best sources of information? The explosion of resources on the Internet is currently the timeliest and most cost-effective means of acquiring fashion news. We now have access to runway-show pictures minutes after the live show occurs, breaking news from across the world, views of international cities and shops, commentary and blogs. We can also get some trend-forecasting subscription services from professional associations on the Internet.

哪里是信息的最佳来源？网上的资源铺天盖地，是目前获取时尚资讯最及时、成本最低的方式。现在我们可以在网上获得几分钟前刚刚发布的时装秀图片和来自世界各地的即时新闻，可以浏览国际大都市和商店的页面，阅读评论和博客。我们还可以在网上获得来自专业协会的趋势预测订阅服务。

Though searching on the Internet is very convenient, designers should continue to travel to visit cities, museums and shows, suppliers and factories. Meeting people face to face; seeing garments on the runway, or in shops; touching fabrics; and observing production are still the richest sources of information, because the experience is live. During this process, personal observation also allows individual interpretation of the information that is not edited by an outside source.

虽然在网上搜索很方便，但设计师还是应该出去参观城市、博物馆和展示会，考察供应商和生产商。与人面对面接触、观看T台或商店的服装、接触面料、考察生产仍然是信息最丰富的来源，因为这些是亲身体验。在此过程中，亲自观察还可以对未经外界编辑的信息进行自我诠释。

If you ask a designer: "What is your inspiration for this collection?" Answers would vary widely, and sometimes may be considered strange. Designers often search for inspiration in the world around them. The source of inspiration is usually a motivating factor for the designer that provides a concept and visual direction.

如果你问一个设计师："你这个系列的灵感是什么"？答案会千差万别，有时甚至很奇怪。设计师经常会从他们周边的世界里寻找灵感。灵感来源往往是带给设计师观念或视觉上引导的刺激性因素。

Inspiration can come from a variety of aspects, such as photographs of fashion shows, fabrics, movies, arts, or personal experiences, etc. Travel is a very rewarding way of searching inspiration that can serve to change ideas and widen viewpoints. Exposure to a new place, different environment and observations can inspire a designer visually, and offer a fresh approach to designing. Travel also allows the opportunity to search and shop in local markets; maybe you can find some funny handmade fabrics, accessories or crafts there. Flea markets and antique fairs are also useful sources for searching inspiration. If large companies have available budgets they will send their designers on research trips, often abroad, to search for inspiration. Designers are armed with a camera or other portable electronics, and are instructed to seek, record and buy anything that might prove useful for the coming season designing.

灵感可以来自很多方面，比如时装发布会照片、面料、电影、艺术或个人经历等。旅行也是寻找灵感的非常有益的方式，有助于转变观念，拓宽视野。去一个新的地方，对不同的环境和事物的观察可以从视觉上启发设计师，带来全新的设计思路。在旅行中还有机会去当地市场上搜寻和购物，也许在那里会发现一些有趣的手工面料、饰品或工艺品。跳蚤市场和古董博览会也是对寻找灵感有益的场所。大公司如果有足够预算的话会派设计师经常到国外考察以便寻求灵感，设计师带着相机或其他便携式电子设备，奉命出去寻找、记录和购买可能对下一季设计有用的任何东西。

Beautiful nature provides designers with limitless inspiration. You can imitate a species' form, color, and texture and apply these elements into fashion design. For instance, in recent years leopard print is very popular and often used to design underwear, scarves and shoes. Another example, designers can learn how to improve the speed of a swimmer by developing swimsuits that mimic the properties of fish that allow them to move quickly.

美丽的大自然给设计师提供了无穷无尽的灵感。你可以模仿一种生物的形态、颜色和纹理，并把这些元素运用到服装设计中。例如，近几年豹纹图案非常流行，经常用来设计内衣、围巾和鞋子。再比如，设计师可以通过研发泳衣获知如何提高游泳者的速度，这些泳衣模仿了鱼儿迅速游动的特性。

Current events such as politics, economy or cultural activities, also provide a huge source of inspiration for designers. For instance, in the 1970s, during women's struggle for equality in the workplace, women's work fashions were patterned after menswear. The navy blue skirted suit with a white blouse and bow tie was a familiar uniform that expressed this struggle. But nowadays women have more power and confidence in the workplace; therefore, they are dressing themselves more beautifully to express feminine strengths.

时事动态如政治、经济或文化活动也给设计师提供了非常多的灵感。例如，20世纪70年代，在女性争取职场平等权利的抗争中，其工作装仿效男装。搭配白衬衫和领结的深蓝色裙套装是表达这场抗争司空见惯的制服。但是现在女性在职场中拥有了更多的权力和自信；她们因此把自己打扮得更加漂亮，以表现女性的优势。

本章小结

- 服装设计有着丰富的含义，服装设计是一个充满挑战与创造力的行业。
- 服装设计的影响因素是多元化的，设计师要了解它们。
- 设计师设计调研和寻找灵感除了利用网络资源以外，还要多出去考察，增加阅历，了解时事动态。

Exercises 课后练习

1. Group exercise 小组练习

Please work in groups of three or four, and discuss which factors will influence fashion design.

三或四人为一个小组，请讨论一下影响服装设计的因素有哪些。

2. Please discuss how to search for fashion design inspiration.

请讨论一下如何寻找服装设计的灵感。

The Evolution of Chinese Clothing
中国服装演变

课题名称： The Evolution of Chinese Clothing

课题内容： 中国服装演变历程

各朝代主要的服装与服饰特征

课题时间： 4课时

教学目的： 让学生了解关于中国服装史的术语名称与英文表达。

教学方式： 结合PPT与音频等多媒体教学课件，以教师课堂讲解为主，学生随
堂练习与小组讨论为辅。

教学要求： 1. 掌握相关词汇。

2. 能够用英文叙述出中国各个朝代主要服装与服饰的名称与特点。

课前（后）准备： 课前预习相关时期的服装史内容，浏览课文；课后记忆单
词，并参照图片描述这一时期的服饰特征。

2.1 Chinese clothing evolution Ⅰ 中国服装演变（一）

2.1.1 Clothing of the Primitive Society (the Paleolithic Age–21st century B.C.) 原始社会服装（旧石器时代—前21世纪）

In the Paleolithic Age, archaeologists did not find anything related to clothing until they discovered some bone-made needles in the Xiaogushanren ruins (around 40000 years ago) and the Upper Cave Man ruins (around 20000 years ago). Each of the bone-made needles had a very smooth sharp needlepoint and a small pinhole. It is possible that early humans used animal's split ligaments as threads to sew clothes. The materials of clothes were probably animal hides, leaves, grass or feathers. These indicated that our ancestors had mastered the skill of sewing and, from then on, a real sense of "clothing" had been appeared. Maybe they wore clothes for warmness, protection, decoration or display.

在旧石器时代，考古学家们曾经没有找到任何与服装相关的事物，直到在"小孤山人"遗址（40000年前左右）和山顶洞人遗址（20000年前左右）中发现了一些骨针，每根骨针都有光滑锋利的针尖和小的针眼。早期人类可能是用动物的韧带作为缝线来缝制衣服，衣服的材料可能是动物的皮革、树叶、草或羽毛。这表明先人们已经掌握了缝制技术；从那时起，真正意义上的"服装"出现了。或许他们穿衣服是为了保暖、防护、装饰或炫耀。

In the ruins of the 6000 year old Yangshao culture of the Neolithic Age, an earthenware bowl was found with textile indentations on the bottom. This textile seemed to be woven from hemp fibers, and there were 10 warps and wefts within each square centimeter. In the ruins of the 4000 year old Liangzhu culture, some silk fragments were found. These showed that the ancient Chinese people had mastered the skills of raising silkworms and weaving silk. The invention of textile technology opened a new page in the splendid history of Chinese clothing culture.

在新石器时代的6000年前的"仰韶"文化遗址中，发现了一个底部有织物凹痕的陶碗。这种织物像是用麻纤维织成，在每平方厘米内有10根经纬纱。在4000年前的"浪渚"文化遗址中，发现了一些丝织物碎片。这说明中国先民此时已经掌握了养蚕和纺丝的技术。纺织技术的发明为灿烂的中国服饰文化翻开了新的一页。

2.1.2　Clothing of the Xia, Shang and Zhou Dynasties (21st century B.C.– 256 B.C.) 夏商周服装（前21世纪—前256）

When human beings entered into the slave society, people's social classes were distinguished clearly and the hierarchy also reflected on clothing. From the emperor down to lower-class people, everyone had to wear the appropriate clothing, headdresses and accessories according to their social stratum, so a person's status could be easily identified from what he or she wore. In China's ancient societies, clothing was closely associated with political governance and rituals, especially in the feudal society.

当人类进入奴隶制社会，人们的社会阶层被明显地区别开来，而这种等级性也反映到服装上。上至天子，下至平民，每个人都要按照其社会阶层穿戴适合的衣裳、冠帽和佩饰，所以一个人的身份很容易通过穿着辨识出来。在中国古代社会，服饰与政治统治还有礼制紧密地联系在一起，特别是在封建社会。

Mianfu was formal attire for emperors, nobles and officials to attend important ceremonies. It was a social institution more than a garment. This social institution which existed for more than 2000 years was passed down from the Western Zhou Dynasty to the Ming Dynasty.

"冕服"是天子、贵族和官员参加重大典礼时穿着的礼服。它不仅仅是一套服装，更是一种社会制度。这种存在了2000多年的社会制度从西周一直沿用到明朝。

Mianfu was composed of a coronet, clothes, fu, a sash, a belt, and shoes. The coronet had yanban (a long board) on the top, which leant forward slightly; this leaning board symbolized the emperor's modesty. Its front brim was round and rear brim was rectangular representing "Round Heaven and Square Earth" (a philosophical opinion of ancient China). Tassels were jade beads bunched by silk threads; they were dangling along the front and rear brims. A jade hairpin was inserted across the coronet to fasten the coronet to the hair bun. Chong'er (two small jade or stone balls) dangled near the ears. Sometimes, when the emperor shook his head, the ball would possibly hit his cheek slightly, to remind him to listen with discretion.

冕服由冕冠、服装、韨❶、丝带、革带和鞋组成。冕冠的顶部有"綖板"（一块木板），稍微向前倾斜，倾斜的綖板象征着天子的谦逊。它前面的边缘是圆形的，后面是方的，代表"天圆地方"（一种中国古代的哲学观）。冕旒是由丝线穿起来的玉珠，在綖板的前后摇晃。玉笄穿过冕冠，把冠体固定在发髻上。"充耳"（两个玉制或石制的小球）垂在耳朵旁边。有时当天子摇头时充耳会轻敲他的脸颊，以提醒他要理性地听取进言。

---

❶ 韨，汉以后称为蔽膝。

The clothes were composed of yi (a jacket) and chang (a skirt). "Shangyixiachang" was a standard man's form of dress for high society members in ancient China. The jacket was black and the skirt was orange red. The black came from the color of the sky at dawn, and the orange red came from the color of the dirt at sunset. This arrangement corresponded to the order of heaven and earth. There were twelve patterns decorated on the clothing.

服装由"衣"和"裳"组成。"上衣下裳"是中国古代上流社会男子的标准着装形式。上衣是黑色,下裳是纁色。黑色来自黎明之天的颜色,下裳来自黄昏之地的颜色,这种搭配对应着上天和下地的顺序。服装上饰有十二章纹样。

Fu was a leather piece hanging in front of the skirt. Although fu did not have any practical use, it could show the wearers' status. Only the high society people were allowed to wear it. There were a silk sash and a leather belt at the waist of the clothes. The silk sash was used to fasten the clothes; the leather belt was used to hang fu and swords.

韨是下裳前面垂挂的一块皮片。虽然韨没什么实际用途,但它可以体现穿着者的身份,只有上流阶层的人才允许佩戴它。服装的腰部系有一条丝带和一条革带。丝带用来固定衣服,革带用来悬挂韨和刀剑。

There were also some strict rules on the wearing of shoes which were matched with mianfu. For different occasions people would wear shoes in different colors. From the very important to less important occasions men would wear red, white and black shoes in sequence. While women would wear black, indigo and red shoes in sequence when wearing formal attire.

与冕服搭配的鞋的穿着也有严格的规定。在不同场合,人们会穿不同颜色的鞋。从非常重要到不太重要的场合。男子会依次穿红色、白色和黑色的鞋子;女子着正装时则依次会穿黑色、青色和红色的鞋子。

In the Spring and Autumn and Warring States periods, shenyi (a special robe) became popular amongst all social classes, and worn by both men and women (Figure 2-1). For the emperor, nobles and officials, shenyi was informal wear or home wear; but for the civilians, shenyi was their formal wear, as well.

在春秋战国时期,"深衣"(一种特殊的长袍)在各个阶层中都很流行,男女都穿(图2-1)。对天子、贵族和官员们来说,深衣是他们的非正式服装和家居服;但对平民来说,深衣还是他们的礼服。

When shenyi was made, the upper part and lower part were tailored separately and seamed together at the waist. This represented respecting the ancestors and succeeding to old traditions (the form of shangyixiachang). Jin (the front piece of a garment) was elongated into a triangle, and then wrapped around the body from front to back and from back to front again. Finally, it was fastened by a silk sash or a leather belt. The collar, jin, the bottom hem, and the cuffs were all added with decorative rims. The lower part of shenyi (like a skirt, but jin excluded) was divided into twelve pieces corresponding to twelve months in a year. There was a vertical line at the back from the top to the bottom implying that the wearer ought to be honest. The bottom hem line was horizontal implying impartial behavior of the wearer.

制作深衣的时候，上半身和下半身分开裁剪，再于腰部缝合。这具有尊重先人，继承传统（上衣下裳制）的含义。"襟"（衣服前片）延长成三角形，从前面绕到背后再绕回前面。最后用丝带或革带固定。领、襟、底摆和袖口都装饰有缘边。深衣的下半部分（裙身，但襟除外）被裁为十二片，对应一年有十二个月。后背从上到下有一条垂直线，表示做人要正直；底摆线是水平的，表示处事要公平。

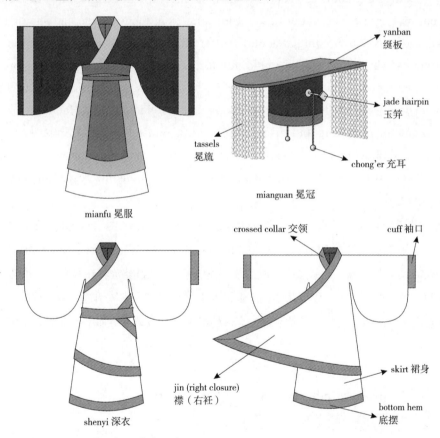

Figure 2-1　Mianfu and shenyi
图2-1　冕服和深衣

The sleeves of shenyi were loose but the cuffs were tight, so the sleeves formed an arc shape when spread. Loose sleeves allowed for flexible movement of the elbow. The collar was a right-slanting crossed collar, which was, the left collar piece covered on the right one and the left collar piece was jointed with jin below. This form was described as right closure, which was one of the traditional Han clothing characteristics. But in minorities' garments, some collars were left-slanting. The sleeve shape of shenyi was round and the neckline of crossed collar was square; this form was also corresponding to that rule—"Round Heaven and Square Earth".

深衣的袖子宽大，袖口缩小，平铺时呈弧形状。宽松的袖子使肘部可以自由活动。领子是右斜向的交领，也就是左领压在右领之上，左领和下面的衣襟相连。这种形式被称为"右衽"，是汉族传统服装的特点之一。但在少数民族服装中，有些领子是左斜向的。深衣袖子的形状是圆的，交领的领线是方的；这种"圆袖方领"的形式也对应了那条法则——"天圆地方"。

During the Warring States period, King Wuling of Zhao State performed a profound reform "wearing hu clothing and shooting on horseback", which was, wearing the clothing of the northwestern ethnic minorities and learning from them to ride and shoot. Their clothing known as hu clothing was quite different from the Han clothing of China's central plains. The northwestern ethnic minorities were nomadic tribes therefore the clothing should be suitable for their lifestyle. They were usually riding on horseback, so they favored wearing short jackets and long pants. Relatively, the Han clothing was inconvenient for mounting and dismounting. King Wuling of Zhao State wanted to introduce the skill of riding and shooting of the ethnic minorities to improve the soldiers' fighting power. To achieve this, he had to reform the Han clothing first, especially chang of the lower part.

战国时期，赵武灵王进行了一次意义深远的改革"胡服骑射"，即穿着西北少数民族服装，学习他们的骑马和箭术。他们的服装被称为"胡服"，不同于中原的汉服。西北少数民族是游牧民族，服装要与其生活方式相适应。他们经常骑在马背上，所以喜欢穿短衣和长裤。相对地，汉族服装不便于上马和下马，赵武灵王欲引进少数民族的骑马和箭术来提高士兵的战斗力。为了达到这一点，他必须要先改革汉族服装，尤其是下裳。

2.2 Chinese clothing evolution II
中国服装演变（二）

2.2.1 Clothing of the Qin and Han Dynasties (221B.C.–220) 秦汉服装（前221—220）

Men of the Qin and Han Dynasties mostly wore a robe which was named pao. Pao had two types according to the front piece: qujupao and zhijupao, and the latter one was more common during this period. Pao evolved from shenyi; qujupao was only popular in the Western Han Dynasty, and then it was gradually replaced by zhijupao. The front piece of zhijupao didn't wrap around the body, but draped straight down to the feet (Figure 2-2). Except for this, the other characteristics were similar to qujupao, such as a crossed collar, right closure and tight sleeve cuffs. Zhijupao was dressed more conveniently, but was only worn as an undergarment before the late Western Han Dynasty.

秦汉男子主要穿"袍"。根据前身的不同袍有两种类型："曲裾袍"和"直裾袍"，后者更常见，袍由深衣发展而来。曲裾袍在西汉时期流行，后来逐渐被直裾袍取代。直裾袍的前身不是包裹住身体，而是一直垂到脚面（图2-2）。除了这一点，其他特点都与曲裾袍类似，如交领、右衽和窄袖口。直裾袍穿起来更加方便，但在西汉末期以前只穿作衬衣。

Zhijupao becoming popular was due to the evolution of pants which were worn inside. Initially the pants were not jointed at the crotch or even had no crotches. The pant legs were separated. The earliest pants were named jingyi; its length was only up to the knee and above the knee was empty. This type of pants was only worn as undergarments and never shown outside. It was considered inelegant if the pants were shown on the outside, so people always wore qujupao or chang outside to cover the incomplete pants. The co-crotch pants were named kun in ancient Chinese language. After kun appeared, qujupao then became unnecessary and was eventually eliminated.

直裾袍变得流行是因为里面穿的裤子的演变。最初裤子的裆部是不缝合的，甚至没有裤裆，两只裤腿是分离的。最早的裤子名为"胫衣"，长度只到膝盖，膝盖以上是空的，这种裤子只穿作内衣从不外露。裤子外露是非常不雅的，所以人们外穿曲裾袍或裳，以遮住不完整的裤子。合裆的裤子在中国古语里叫"裈"。裈出现以后，曲裾袍就变得多余而最终被淘汰了。

The clothing institution was very strict in the Han Dynasty. People's hierarchy and ranks could be distinguished by headdresses and accessories. The headdress worn by civil officials and military officials were different. The civil officials' headdress was known as liangguan, which had vertical strips on the upper part; and the quantity of strips distinguished the ranks of officials. The military officials' headdress was known as heguan, which was decorated with tail feathers of heniao (a bird in legend) on the top. Changguan was said to be designed by Liu Bang, the first emperor of the Western Han Dynasty, which had a tilted bamboo board on the top; it could be worn by emperors and officials. Xiezhiguan (xiezhi was a unicorn in legend) was worn by judges. Before people wore the headdress, they often wore ze first. Ze was a kind of kerchief to fasten the scattered hair; it could be worn by both the emperor and the civilian.

汉代的服装制度非常严格，人们的阶层和品级可以通过冠帽和饰品区别开来。文官和武官的冠帽不同。文官的冠叫"梁冠"，冠的上方有垂直的线条，线条的数量可以区分官员的品级；武官的冠叫"鹖冠"，顶部饰有"鹖鸟"（一种传说中的鸟）的尾羽。"长冠"据说是由西汉的第一个皇帝刘邦创建的，长冠的上方带有一个翘起的竹皮板，皇帝和官员都可戴用。"獬豸冠"（"獬豸"是传说中的独角兽）是法官戴的。戴冠之前要先戴"帻"。帻是束住散发的发巾，从皇帝到平民都可戴用。

In the Qin and Han Dynasties the high society women still wore shenyi as their formal attire. As during pre-Qin periods its elongated front piece wrapped the body, but the tail became longer than before. Sometimes women added a special train named pianzhuyuan to the bottom hem, just like a blooming trumpet. Women were graceful with their feet veiled. And the collar was lower than before like a V shape, so that the collars of undergarments would be shown; this attire was named triple-layer collar.

在秦汉时期上流社会的女子仍以穿深衣为贵。和先秦时期一样，衣襟延长包裹住身体，但是裙裾比以前加长了。有时还会在裙摆处加一块叫作"偏诸缘"的拖裾，好像盛开的喇叭花。长裙掩足，使女子婀娜多姿。衣领也比以前低了，呈V形，于是里面所穿禅衣的领子露了出来；这种装束叫作"三重衣"。

For the common women, they often wore ruqun in their daily life. Ru was a short jacket; qun was a long skirt. Ru also had a crossed collar; the length was below the waistline. The skirt length would veil the feet. The matching of ru and skirt was the most popular style for the common women in ancient China.

对普通妇女来说，她们在日常生活中常穿"襦裙"。"襦"是一件短上衣；"裙"是长裙。襦也有交领，长度到腰线以下；裙的长度要盖住脚面。襦裙搭配是中国古代普通妇女最普遍的装束。

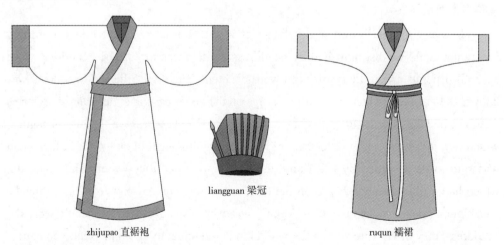

zhijupao 直裾袍

liangguan 梁冠

ruqun 襦裙

Figure 2-2　Clothing of the Han Dynasty
图2-2　汉代服装

2.2.2　Clothing of the Wei, Jin, Northern and Southern Dynasties (220– 589) 魏晋南北朝服装（220—589）

Loose-fitting clothing prevailed in the Wei and Jin Dynasties because the scholars of that period were pursuing a carefree wearing style. Most men of this period wore shan. Shan was a kind of loose robe with open sleeves; lined or unlined, having front opening or right slanting front. If it was front-opening, the opposite front pieces were fastened with a sash sometimes. White was the most popular color for shan. The labor class men still wore short jackets along with pants.

魏晋时期流行宽衣博带，因为那时的文人追求自然随性的穿着风格。男子主要穿"衫"。衫是一种宽松大袖的长袍；有单、夹二式，对襟或右衽。如果是对襟，衣襟有时用带子系住。衫以白色为尚。劳动阶层的男子仍然穿短衣长裤。

The officials in the Wei and Jin Dynasties often wore qishalongguan (Figure 2-3), a black cylindrical headdress. It was made of black gauze on which was painted a layer of black lacquer to make it firm. It was extended downward encasing the ears and back of the head, finally fastened by ribbons below the chin.

魏晋时期的官员经常佩戴"漆纱笼冠"（图2-3），一种黑色筒状的冠。它用黑纱制成，外面涂了层漆水，使之坚挺。它向下延长包裹住耳朵和后脑勺，最后用带子系于颌下。

Noble women still wore shenyi as their formal clothing. But shenyi of this period was quite different from before. The lower part of shenyi was composed of several triangular pieces which were like banners. These pieces and some silk ribbons stretching from the pieces created a layered effect, and lent rhythm to the women's movement.

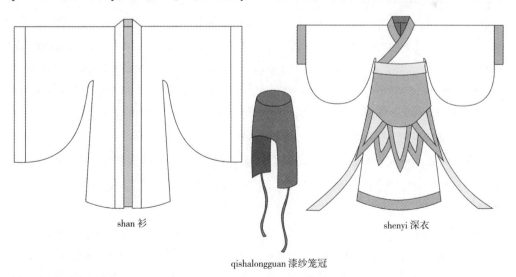

shan 衫

qishalongguan 漆纱笼冠

shenyi 深衣

Figure 2-3　Clothing of the Wei, Jin, Northern and Southern Dynasties

图2-3　魏晋南北朝服装

贵族女子仍然以深衣作为她们的正式着装。但这时期的深衣和以前相比差别明显。深衣的下半部分由一些旗帜似的三角形裁片组成，这些裁片和从裁片里伸出来的数条丝带形成了一种层叠相间的效果，给女子的行动增添了许多韵律。

Common women mainly wore ruqun. Originally the loose-fitting style prevailed; later the jacket became narrow because of the influence of the northern ethnic minorities. This style is described as "narrow upper, loose lower". Sometimes women preferred wearing an overskirt on the long skirt, which could make them more elegant.

普通女子主要穿襦裙。起初襦裙流行宽博的风格；后来受北方少数民族的影响，上衣变得紧窄，这种风格被形容为"上俭下丰"。有时妇女还喜欢在长裙外面加一条小罩裙，可以让她们更加优雅。

Women in the Wei and Jin Dynasties paid a lot of attention to makeup. E'huang was a kind of popular makeup of this period. It dyed women's forehead light yellow to make them beautiful. This makeup was derived from Buddhism, as women got the inspiration from the golden yellow Buddha in temples. Huadian was another fashionable ornament, which was worn on the forehead between eyebrows. It could be made of thin paper, shell, gold foil, jade flake or pearl, etc. It varied in shapes with fine workmanship, and was pasted by special glue.

魏晋女子非常重视面妆。"额黄"是这一时期的流行妆容。它是把女子的额头染成淡黄色来美化容貌。这种妆容来自佛教，因为妇女们是从寺庙中金黄色的佛像那里取得灵感。"花钿"是另外一种流行的妆饰，点于眉心之间。它可以用薄纸片、贝壳、金箔、玉片或珍珠等材料做成，造型各异，做工精细，并用一种特制的胶粘贴。

2.2.3　Clothing of the Tang Dynasty (618–907) 唐代服装（618—907）

In terms of cultural and economic development, the Tang Dynasty was undoubtedly a peak period in the development of human civilization in ancient China. The Tang government was tolerant and appreciative of religion, art and culture from the outside world, which profoundly influenced the clothing of the Tang, especially women's clothing, making it unique and attractive.

在文化和经济发展方面，唐代无疑是中国古代文明发展的鼎盛时期。唐王朝对外来宗教、艺术和文化持包容和欣赏的态度，这些对唐朝服装特别是女装影响深远，使其独特而富有魅力。

Men of the Tang Dynasty mainly wore round-neck robes. The robe had right slanting front, a round neck, narrow sleeves, and slits on both sides; some of the characteristics were borrowed from the ethnic minorities' clothing. Some robes had two jointed pieces named henglan at the bottom front and back, and henglan's fabric was the same as the robe. The use of henglan implied that the robe succeeded to the traditional form (the form of shangyixiachang). Men wore a leather belt and black leather boots to go with the robe. The officials' ranks were distinguished by colors of robes; the color sequence from high to low ranks was purple, red, green and cyan. Furthermore, from the emperor of Tang Gaozong the use of yellow was monopolized by the royals; no one else could wear yellow if not be permitted.

唐代的男子主要穿圆领袍服。其为右衽、圆领、窄袖，两侧有开衩；其中一些特点是从少数民族服装中借用来的。有些圆领袍服的前后底端有两块拼接的裁片，叫作"横襕"，面料和袍身一样，横襕的使用意味着对传统形式（上衣下裳制）的继承。男子穿袍服要搭配革带和黑色的革靴。官员的品级以袍服的颜色来区分，按照等级从高到低颜色依次为紫、绯、绿、青。此外，从唐高宗开始，黄色的使用被皇族人垄断，未经允许任何人不能穿黄色。

Officials and civilian men in the Tang Dynasty wore futou on their heads. Futou was a kerchief wrapping around men's heads with two ribbons on the back. The early futou was simple; later it was filled with a pad made of wood, moss, silk or leather to make various forms. The ribbons were initially soft, and then became stiff with some metal threads inserted in them.

唐代官员和平民男子头戴"幞头"。幞头是包在头上、后面垂有两根带子的发巾。早期的幞头比较简单，后来在里面加了一个用木头、苔藓、丝织物或皮革做成的衬垫物，以撑出各种形状。带子最初比较柔软，在里面加了一些金属丝后就变得硬挺了。

Tang women's clothing can be classified in three categories: ruqun, male garments and hu-clothing (Figure 2-4). Ruqun was a traditional type, which was made up of ru or a long gown and a long skirt. Ruqun in Tang Dynasty was very distinctive, reflecting the open social atmosphere. The top jacket collar was opened up as far as exposing the cleavage between the breasts. This was unprecedented in previous dynasties, in which women had to cover their entire body. The long gown was very thin and light, beneath which women's shoulders and the back could be faintly seen. The skirt was very long and the waistband was lifted to above or below the breasts. This style did not appear in other dynasties. A unique skirt known as pomegranate-red skirt was very popular in this period; it had vivid color that was as red as the blooming pomegranate flower in May.

唐代的女装可以分为三种类型：襦裙装、男装和胡服装（图2-4）。襦裙装是一种传统的样式，由襦或长衫和长裙组成。唐代的襦裙装很有特色，反映出开放的社会风气，上衣的领子开得很大，直到露出乳沟。这在以前的朝代中是未曾有过的，之前的女子要遮住整个身体。长衫很轻薄，使女子的肩部和后背若隐若现。裙子很长，裙腰提高到胸上或胸下，这在其他朝代都没有出现过。这一时期还流行一种独特的裙子"石榴裙"，它色彩艳丽，就像五月盛开的石榴花一样红。

When wearing ruqun, women liked matching banbi and pibo. Banbi was a short coat with short or half sleeves, worn over the long sleeved jacket. Pibo was made of silk, draped over the arms from back to front. It had two types: the wide and short type worn like a cape; the narrow and long type worn like a ribbon.

穿襦裙时，女子喜欢搭配"半臂"和"披帛"。半臂是件短袖或半袖的短上衣，穿于长袖上衣的外面。披帛是由丝织物做成的，从后向前绕过手臂。它有两种造型：又宽又短型，穿起来像披肩；又细又长型，穿着后像飘带。

A woman wearing male garments was very rare in other dynasties, but it was very common in the Tang Dynasty. Women could wear full sets of men's clothing, including futou, a round-neck robe, boots and a leather belt, or even could wear their husbands' garbs when going out. In the Tang Dynasty hu-clothing not only referred to the northern nomadic tribes clothing, but included the foreign costume that came from the Silk Road. The main style of hu-clothing was robes with narrow sleeves and turn-down collars, matching with pants and leather belts.

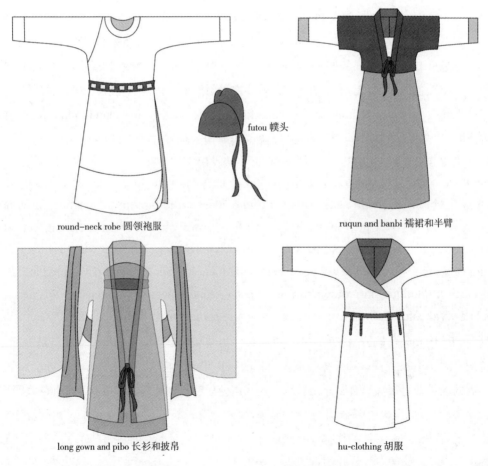

futou 幞头

round-neck robe 圆领袍服

ruqun and banbi 襦裙和半臂

long gown and pibo 长衫和披帛

hu-clothing 胡服

Figure 2-4　Clothing of the Tang Dynasty
图2-4　唐代服装

女子穿男装在其他朝代非常罕见，但在唐朝司空见惯。女人们可以穿戴一整套男子的装束，包括幞头、圆领袍、长靴和革带，甚至可以穿着她们丈夫的衣服出门。在唐代，胡服不仅指北方游牧部落的服装，还包括来自丝绸之路的异国服饰。胡服的主要类型是窄袖、翻领的长袍搭配长裤和革带。

There was a large variety of hairstyles in this period, most of them were high buns and extravagant in style. The facial makeup was also diverse. Huadian of the Wei and Jin Dynasties was remained and developed. It was decorated not only on the forehead, but also on the temple, cheeks and both sides near the mouth. The shape would be like coins, birds, or flowers and so on. A kind of makeup made women look sad as they painted slanted eyebrows and black lips; this was called "weeping makeup".

这个时期有大量的发型式样，多数是高发髻，风格奢华。面妆也是多种多样。魏晋时期的花钿被继续使用和创新，它不仅饰于额头之上，也可以装饰于太阳穴、脸颊和嘴角的两边，其形状有铜钱形、鸟形和花形等。有种妆容让女人看起来很伤心，因为她们画有八字眉和黑色的嘴唇；这被称为"泣妆"。

2.3 Chinese clothing evolution Ⅲ
中国服装演变（三）

2.3.1　Clothing of the Song Dynasty (960–1279) 宋代服装（960—1279）

Officials of the Song Dynasty usually wore lanshan. Most features of lanshan were similar to the round-neck robe of the Tang Dynasty; they both had henglan on the bottom. However, lanshan was loose fitting and the sleeves were broad. The officials' ranks were also differentiated by colors as during the Tang. Officials' headdress was mostly futou (Figure 2-5), but differed from that of the Tang Dynasty. It had two straight sticks stretching on the sides; it is said that these were used to prevent officials whispering to each other when in court.The common men generally wore robes, shan, short jackets, pants and beizi. Most colors of their clothing were dull and dim.

宋代官员通常穿"襕衫"。襕衫的多数特征和唐代圆领袍服相似，底摆都有横襕。但襕衫袍身宽松，袖子宽大。和唐代一样，官员的品级也可以通过颜色区分。官员的首服主要是幞头（图2-5），但与唐代不同。它的两侧伸出两个展角，据说是为防止官员在上朝时交头接耳而用的。普通男子一般穿袍、衫、短衣、长裤和褙子，这些衣服的颜色大多比较灰暗。

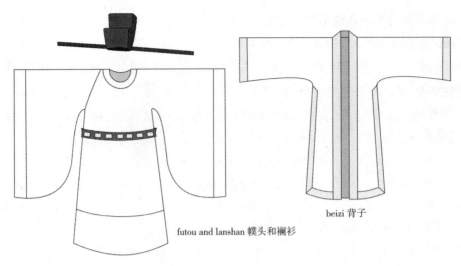

futou and lanshan 幞头和襕衫

beizi 背子

Figure 2-5　Clothing of the Song Dynasty
图2-5　宋代服装

Women of the Song Dynasty regardless of their social classes mostly wore beizi. Beizi could be worn by both sexes but mostly by women. It was a front-opening overcoat; the length was above the knee, below the knee or down to the ankle. The sleeves could be either broad or narrow. There were two side slits reaching as high as the armpits, or none at all. Compared to the extravagant style of the Tang garments, Song people preferred reserved elegance. This was because Song clothing was greatly influenced by a prevalent ideology of the time—Neo-Confucianism. In addition, most Song women bound their feet. From then on this unhealthy custom influenced women's life for more than a thousand years.

宋代女子不管哪个阶层都常穿褙子。男女都可以穿褙子，但以女性为多。它是对襟式外衣，长度在膝上、膝下或及踝。袖子可宽可窄。两侧有开至腋下的衩口或没有开衩。相对于唐代服装奢侈张扬的风格，宋朝人偏爱内敛的优雅，这是因为宋代服装受当时流行的意识形态——理学的影响很大。此外，宋代女子大多裹脚。从那时起这种不健康的习俗影响了女性的生活一千多年。

2.3.2　Clothing of the Liao, Jin, and Yuan Dynasties (907–1368) 辽金元服装（907—1368）

The regimes of Liao, Jin and Yuan were established by northern ethnic minorities. They coexisted with the Han national political regime. The clothing of the Liao, Jin and Yuan Dynasties was much influenced by that of the Han nationality. These ethnic minorities' garments mostly were robes with a round neck, narrow sleeves and left closure; men wore robes with pants and women wore with skirts. Men's hairstyle was named kun hair, which was different from Han men's. They shaved off their hair on the top and back of the head, but only reserved few forelocks and hair on the temples; while Han men always reserved all of their hair. Noble women of the Yuan Dynasty often wore a unique headdress named guguguan, which was tall and thin with bamboo or wood making the frame (Figure 2-6). It was said that the higher the social status, the taller the headdress.

kun hair 髡发

guguguan 顾姑冠

Figure 2-6   Kun hair and guguguan
图2-6   髡发和顾姑冠

辽、金、元政权是由北方少数民族建立的，他们与汉族政权共存。辽金元时期的服装深受汉族服装的影响。这些少数民族的服装主要是圆领、窄袖、左衽的长袍；男子穿长袍配长裤，女子配长裙。男子的发式名为"髡发"，不同于汉族男子的发式。他们剃去了头顶和脑后的头发，只保留少量的额发和鬓发；而汉人却一直蓄发。元代贵族女子常戴一种叫"顾姑冠"的独特冠饰，又高又细，用竹子或木头制成骨架（图2-6）。据说社会地位越高，顾姑冠也就越高。

2.3.3   Clothing of the Ming Dynasty (1368–1644) 明代服装（1368—1644）

Official uniforms of the Han nationality were resumed during the Ming Dynasty and were worn more strictly than in the previous periods. Emperors wore bright yellow dragon robes with dragon patterns embroidered on the front, back and shoulders. They wore yishanguan and a jade belt to go with the robe. Officials mostly wore patch robes which were decorated with buzi (an ornamental patch) on the chest and back. Different robe colors showed different ranks from high to low being red, cyan and green. Each rank of the robes was embroidered with different patterns on the patch; the civil officials used birds and military officials used beasts. Thus, the officials' categories and ranks were clearly differentiated. Wearing a patch robe would match a black gauze hat, a belt and black boots. This attire often appeared in Beijing opera as well. The common men usually wore straight robes which had dark rims on the collar, cuffs and bottom hems. The colors mostly were white, light blue and black. Men would wear a scholar hat which was folded by a black kerchief to match with the robe.

汉族的官服在明朝重新穿用，并且比以前穿着更加严格。皇帝穿明黄色的龙袍，在前胸、后背和肩部绣有龙纹。穿龙袍要搭配"翼善冠"和玉带。官员们主要穿前胸后背饰有"补子"（装饰性补裰）的补服。不同的袍服颜色表示不同的品级，从高到低分别是绯、青、绿。每一品级在补子上绣有不同的花纹，文官绣禽，武官绣兽，官员的类别和品级可以明显地区别开。穿补服要搭配乌纱帽、革带和黑色的靴子。这种装束也经常出现在京剧中。普通男子经常穿领、袖和底摆饰有深色镶边的直身袍。其颜色主要是白、浅蓝和黑色。男子穿直身袍会搭配用黑色头巾折叠成的儒巾。

Empresses wore red long gowns with di (a bird in legend) pattern on it, matching with phoenix crowns and peizi (a long ornamental strip draped around the neck). This attire was depicted as "fengguanxiapei". The common women wore beizi, ruqun, bijia and shuitian dress (Figure 2-7). Bijia was a long sleeveless waistcoat worn over beizi, and draped all the way down to below the knee. For Ming women a slender figure was the ideal of beauty. Bijia helped create a visual impression of slenderness. Shuitian dress was a typical style of Ming women's garments; it was created by labor women. Its style was similar as beizi but was sewn with many different fabric pieces, which made it colorful and beautiful.

皇后穿绣有翟（传说中的鸟）纹的红色长袍，搭配凤冠和帔子（垂挂于颈间的装饰长条），这种装束被形容为"凤冠霞帔"。普通女子穿褙子、襦裙、"比甲"和"水田衣"（图2-7）。比甲是穿在褙子外面的无袖长马甲，一直垂到膝盖以下。对明代女性来说苗条的身材是美的典范。比甲有助于塑造苗条的形象。水田衣是一种典型的明代女装，它由劳动妇女创造。它的造型跟褙子相似，但是由多块不同的面料碎片缝制而成，使其色彩斑斓而美观。

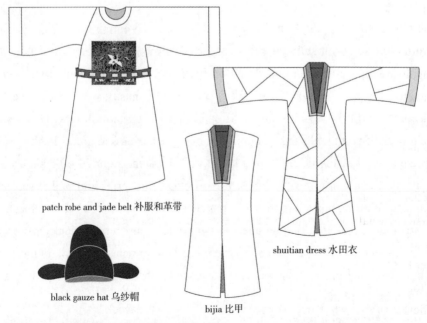

patch robe and jade belt 补服和革带

black gauze hat 乌纱帽

bijia 比甲

shuitian dress 水田衣

Figure 2-7 Clothing of the Ming Dynasty
图2-7　明代服装

2.4 Chinese clothing evolution IV
中国服装演变（四）

2.4.1 Clothing of the Qing Dynasty (1636–1912) 清代服装（1636–1912）

The Qing Dynasty was established by the Manchu people. After they took control, they first reformed the Han national clothing. The Qing government strictly forbade Han people to wear Han traditional clothing and headdress. Han people had to wear Manchu's garb and men also had to shave off their forelocks, only leaving a long braid at the back. This irritated Han people severely and triggered their strong resistance. The Qing government compromised at last to ease the crisis. Therefore, Han women in the Qing Dynasty were permitted to wear Han traditional clothing, but men were not.

清朝是由满族人建立的。他们掌握政权以后，首先对汉族服装进行了改革。清政府严厉禁止汉人穿戴汉族传统的服饰和首服。汉人必须穿满族服装，男子还要剃掉他们的额发，只在身后留一条长辫子，这严重激怒了汉人并引起他们强烈的反抗。清政府最终采取了折中的办法来平息危机。因此，清朝的汉族女子允许穿汉族传统服装，但男子不可。

The Qing Dynasty had the most complicated clothing institution in Chinese history. As for the emperor's court dress (Figure 2-8), it was a bright yellow dragon robe embroidered with dragon patterns. Its distinctive sleeve cuffs were named horse-hoof sleeves that could cover the hands in cold days when turned down. According to different seasons, the emperor would wear a summer court dress or a winter court dress. The winter court dress added precious fur rims to the cuffs, the slanting placket and hems. When wearing a court dress, the emperor would also wear a summer court hat or a winter court hat, a cape, a string of court beads, a court belt and court boots.

清朝具有中国历史上最复杂的服装制度。就皇帝的朝服（图2-8）来说，是一件绣有龙纹的亮黄色龙袍。它独特的袖口名为"马蹄袖"，在天气冷的时候可以翻下来罩住双手。根据季节不同，皇帝会穿夏朝服或冬朝服。冬朝服在袖口、斜襟和下摆处添加了珍贵的皮毛镶边。穿朝服时，皇帝还要搭配夏朝冠或冬朝冠、披领、朝珠、朝带和朝靴。

When officials were in court, they mostly wore jifu (Figure 2-9). Jifu was composed of a python robe, a patch coat (a coat embroidered with buzi), jifu hat (contained the summer hat and the winter hat), a string of court beads, a cape and black boots etc. There was a top bead on the top of the hat, and yuling at the rear of the hat; they both signified the officials' ranks. Some hats' yuling was made of the peacock's tail feather; the pattern of the feather was known as yan. The python pattern was slightly different from the dragon

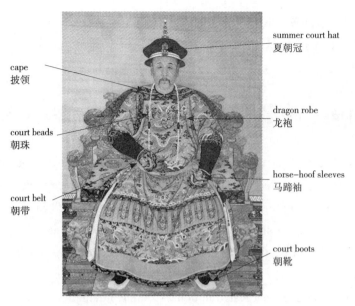

cape
披领

court beads
朝珠

court belt
朝带

summer court hat
夏朝冠

dragon robe
龙袍

horse-hoof sleeves
马蹄袖

court boots
朝靴

Figure 2-8　The emperor's court dress
图2-8　皇帝朝服

pattern. From high ranking to low ranking officials, the number of pythons on their robes was respectively nine, eight and five. The patch patterns were a little different from those of the Ming Dynasty. The patch had a round shape and a square shape; the round shape patch was used by the royals and the square shape by common officials.

官员上朝的时候，通常穿"吉服"（图2-9）。吉服由"蟒袍"、补服（绣有补子的外褂）、吉服冠（包括暖帽和凉帽）、朝珠、披领和黑色的靴子等组成。冠的顶部有顶珠，后面有"羽翎"；都代表着官员的级别。有些羽翎用孔雀的尾羽制作；其花纹被称为"眼"。蟒纹和龙纹差别微妙。从高级官员到低级官员，其蟒袍上蟒纹的数量分别是九条、八条和五条。补子的图案和明朝略有区别。补子有圆形和方形两种；圆形补子是皇族人用的，方形补子是一般官员用的。

On informal occasions, the royals, officials and common people usually wore robes, mandarin jackets, waistcoats, pants, and so on. A type of skullcap was very popular in the Qing Dynasty and the Republican period. Its shape was like a half watermelon shell; hence it got the name "melon-shell cap".

在非正式场合，皇族人、官员和平民通常穿长袍、马褂、马甲和长裤等。有种无檐便帽在清朝和民国时期都非常流行。它的形状像半个西瓜皮，因此得名"瓜皮帽"。

The empress's court dress (Figure 2-10) was somewhat similar to the emperor's, but more complicated, especially the accessories. From inside to outside the empress would wear a court skirt, a court robe and a court waistcoat. When wearing the court dress, the empress also wore a summer court hat or a winter court hat, beneath which they wore

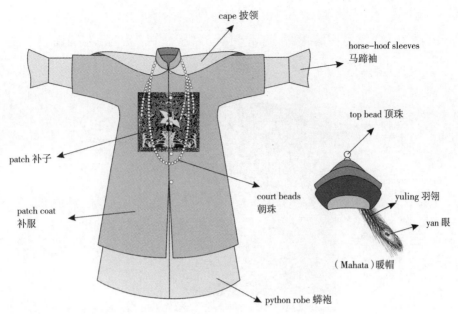

cape 披领

horse-hoof sleeves 马蹄袖

top bead 顶珠

patch 补子

court beads 朝珠

yuling 羽翎

patch coat 补服

yan 眼

（Mahata）暖帽

python robe 蟒袍

Figure 2-9　Jifu of the officials

图2-9　官员吉服

jinyue (a golden head hoop). She wore two strings of coral beads on the court waistcoat, and then wore a cape, another string of court beads and lingyue (a golden neck ring) in sequence. An ornamental ribbon caishui was fastened to one of the court waistcoat buttons; its colors and patterns would signify the wearer's status.

皇后的朝服（图2-10）和皇帝的有些相似，但更复杂，特别是配饰。皇后从内到外穿朝裙、朝袍和朝褂。穿朝服时，皇后还会戴夏朝冠或冬朝冠，朝冠的下面戴"金约"（金制头箍）。在朝褂的上面先戴两盘珊瑚珠，然后依次在上面穿披领，再戴一条朝珠，戴"领约"（金制项圈）。装饰性的绸带"彩帨"系在朝褂的一颗纽扣上，上面的颜色和纹样可以显示穿着者的身份。

Manchu women wore long gowns in their daily life, which were usually called cheongsam. It was right-slanting and decorated with intricate and exquisite embroideries. Cheongsam (Figure 2-11) had no collar, so women always matched a white scarf with it. Some of the cheongsam had two side slits on the gown that were convenient for Manchu women to ride horses before they settled in central China. They wore pants under the cheongsam and a short waistcoat over it.

满族女子在日常生活中穿长袍，通常叫"长衫"，其为右衽，且装饰有复杂、精美的刺绣。长衫（图2-11）没有领子，所以穿着时要搭配一条白色的领巾。有些长衫的两侧有开衩，是在满族入主中原以前方便满族妇女骑马用的。她们在长衫的里面穿长裤，外面穿短坎肩。

Manchu women's hairstyle and shoes were very distinctive. They combed their hair

winter court hat
冬朝冠

jingyue
金约

lingyue
领约

coral beads
珊瑚珠

court waistcoat
朝褂

court robe
朝袍

caishui 彩帨

Figure 2-10　The empress's court dress
图2-10　皇后朝服

into a horizontal chignon at the back supported by bianfang (a metal board). In the late Qing Dynasty, court women favored wearing qitou on the top of the head, which was a high black board with artificial flowers, jewels and tassels on it. Manchu women did not have the custom of feet binding. Their shoes had a high heel at the middle of the sole; these wooden heels were shaped like flowerpots or horse hooves.

满族女子的发式和鞋很有特色。她们把头发在后面梳成一个水平的发髻，用"扁方"（金属板）支撑。在晚清时，宫廷女子喜欢在头顶戴"旗头"，这是一块高高的黑色板子，饰有人造花、珠宝和流苏。满族女子没有裹脚的习俗。她们的鞋在鞋底中央有一个高跟，这些木制的鞋跟形状像花盆或马蹄。

Han women of the Qing Dynasty still maintained the Han wearing style; they usually wore ao and a skirt. Ao was a loose-fitting jacket with dropped shoulder sleeves. Later due to the influence of the Manchu's clothing, wearing pants to accompany the ao became more popular. As with Manchu clothing, Han clothing also had intricate embroidery decoration. Han women still maintained the old custom of binding feet. Having a pair of small, pointed feet was considered to be an essential feature for a beautiful woman.

清朝的汉族女子仍保留着汉族的穿着风格，她们通常穿"袄"裙。袄是一件宽松的带有落肩袖的上衣。后来由于受满族服饰的影响，袄和长裤搭配变得更加流行。和满族服装一样，汉族服装也有繁复精美的刺绣装饰。汉族女子仍保留着旧的裹脚习俗，拥有一双尖尖的小脚被看作是一个美女的基本特质。

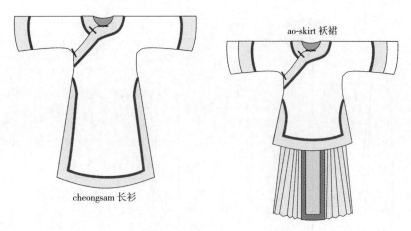

ao-skirt 袄裙

cheongsam 长衫

Figure 2-11   Women's clothing of the Qing Dynasty
图2-11　清朝女装

2.4.2   Clothing of the Republican period (1912–1949) 民国时期服装（1912—1949）

During the Republican period, influenced by European and American fashion, the traditional clothing style and forms of the Qing Dynasty gradually changed. The hierarchical distinctions of clothing disappeared. Besides traditional robes and mandarin jackets, young men also wore student uniform (Figure 2-12). It was brought in from Japan and followed the western tailored suit style. It had a stand collar; on the front pieces of the jacket there were three welt pockets; one was on the left chest and two were on the lower part. The modern typical Chinese men's clothing—the Sun Yat-sen uniform was adopted from it.

民国时期，受欧美时尚的影响，清朝传统的服装风格和样式逐渐改变，服装的阶级区别消失了。除了传统的长袍马褂外，青年男子还穿"学生装"（图2-12），它是从日本引进，并沿用了西方的西服样式。学生装为立领，前身有三个挖袋，一个在左前胸，另两个在下面。典型的现代中式男装——"中山装"便是由它演变而来。

The Sun Yat-sen uniform was very popular in the Republican period and after 1949. It was named after Sun Yat-sen—the founder of the Republic of China. This uniform had a turn-down collar, five buttons and four patch pockets on the front, three buttons near the sleeve opening. The Sun Yat-sen uniform was very practical; it also conforms to the aesthetic preference of Chinese people, because it possesses the appearance of symmetry, solemnity and preciseness.

中山装在民国时期和1949年以后都很流行。它是以"民国之父"孙中山先生的名字命名的。它带有翻领，前身五粒纽扣和四个贴袋，袖口三粒纽扣。中山装非常实用，它也符合中国人的审美偏好，因为它具有对称、庄重和严谨的外观。

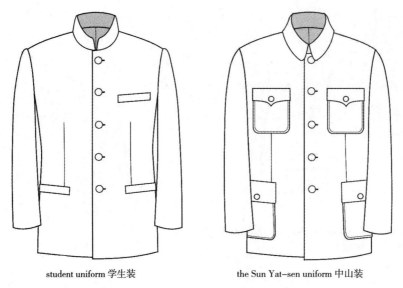

student uniform 学生装　　　　the Sun Yat-sen uniform 中山装

Figure 2-12　Men's clothing of the Republican period

图2-12　民国时期男装

Cheongsam was popular among women but was different from that of the Qing Dynasty. This type which was named improved cheongsam was later considered a form of Chinese classic traditional clothing. It has a mandarin collar, the right closure, edge binding, and two side slits. Compared to the previous traditional clothing which did not highlight women's waistline, it is very form-fitting. Thus, cheongsam was able to show the beauty of the female figure. In the 1920s its outline was similar to that of the Qing Dynasty; they were both loose fitting. However, the embroideries were decreased and the sleeves and skirts were shortened; bell-shaped sleeves prevailed. In the 1930s cheongsam became form-fitting. Most cheongsam in the 1930s had long bottom hems, which were down to the ankles or even the floor. When wearing cheongsam, elegant ladies would also match stockings and high heels. In the 1940s because of the influence of anti-Japanese war, the economy declined. The length of cheongsam sleeves and skirts were shortened again. Knee-length styles were common and the fabrics were plainer.

旗袍很受女性欢迎，但与清朝时的不同。这种被称为"改良旗袍"的款式日后成为中式经典的传统服装。旗袍为立领、右衽、绲边，有两条侧开衩。与以往不凸显女性腰身的传统服装相比，它非常合体，因此，改良旗袍能够展现女性的身材美。在20世纪20年代，它的造型与清朝时期的相似，都是宽松式；但是刺绣减少了，袖长和裙长缩短，钟型袖流行。到了20世纪30年代旗袍变得合身，大多数20世纪30年代的旗袍底摆很长，到脚踝甚至地面；穿旗袍时，优雅的女士还会搭配长筒袜和高跟鞋。20世纪40年代因受抗日战争的影响，经济衰退，旗袍的袖长和裙长又变短了，齐膝裙非常普遍，面料也更加朴素。

2.4.3 Clothing of the People's Republic of China period (1949–1990) 中华人民共和国服装（1949–1990）

After the People's Republic of China was founded in 1949, the aesthetic and style of clothing were changed considerably. Except for the Sun Yat-sen uniform, most clothing styles were abandoned. People preferred to dress themselves like labor classes. Chinese-style short coats and loose–fitting pants became part of this new fashion. The clothing colors were very monotonous; blue, gray and black were the mainstream. Women occasionally wore printed cloth cotton coats or floral dresses, but the colors and patterns were still plain. In the 1950s because of the close Sino-Soviet relations, the Lenin coat (Figure 2-13) became prevalent; it was usually worn by some female cadres.

Figure 2–13   Lenin coat
图2-13   列宁装

1949年中华人民共和国成立以后，服装的审美和风格发生了巨大的变化。除了中山装，大部分的服装样式都被丢弃了，人们倾向于把自己穿得像劳动人民，中式上装和宽松裤成为流行服装。服装颜色非常单一，蓝、灰、黑是主体色。妇女们偶尔穿印花的棉袄或碎花连衣裙，但颜色和图案仍很朴素。20世纪50年代由于中苏关系密切，列宁装（图2-13）流行，它通常被女干部穿着。

During the 1960s and 1970s, the service dress became popular. It was composed of olive green jackets and pants, matching with liberation caps and army satchels. This style did not end until the reform and opening-up policy began in 1978. After the 1980s along with the opening of people's ideas, clothing became diversified and many international clothing brands were introduced. Nowadays the clothing variety is very abundant and people's dressing is more and more individualized.

在20世纪60年代到20世纪70年代，军便装成为流行。它由橄榄绿的上衣和裤子组成，搭配解放帽和军用背包，这种风格直到1978年改革开放以后才结束。20世纪80年代以后随着人们观念的开放，服装变得多元化，许多国际服装品牌被引入中国。现如今服装的种类非常丰富，人们的穿着也越来越个性化。

■ 中国古代服装体现着严格的等级性和制度性。

■ 在中国古代，服饰是一个人的名片，服装、冠帽、配饰要与穿着者的身份相适应。

■ 很多服装或配饰体现着传统的礼仪习惯或社会风俗，服饰是文化的载体与缩影。

课后练习

1. Please retell the characteristics of main clothing, headdresses and accessories in each dynasty.

请复述每个朝代主要的服装、冠帽和配饰的特征。

2. Group exercise 小组练习

Please work in groups of three or four students, and discuss and exchange your opinions about the clothing of some dynasties.

三或四人为一个小组，请对某些朝代的服装进行讨论和交流。

The Evolution of European Costume
欧洲服装演变

课题名称： The Evolution of European Costume

课题内容： 欧洲服装演变的进程

各个时期服装的风格与典型的款式特征

课题时间： 4课时

教学目的： 让学生了解关于欧洲服装史的术语名称与英文表达。

教学方式： 结合PPT与音频等多媒体教学课件，以教师课堂讲解为主，学生随

堂练习与小组讨论为辅。

教学要求： 1. 掌握相关词汇。

2. 能够用英文叙述出欧洲各个时期服装风格的特征与基本款式特点。

课前（后）准备： 课前预习相关时期的服装史内容，浏览课文；课后记忆单

词，并参照图片描述这一时期的服饰特征。

3.1 European costume evolution Ⅰ 欧洲服装演变（一）

3.1.1　Ancient Egyptian costume 古埃及服装

As society evolved, clothing had become a way to show an individual's rights, dignity and wealth. Throughout the 3000 years of Egyptian history, slaves, peasants and other lower class people only wore the simplest of garments, which showed these people's relatively insignificant social status.

随着社会的发展，服装成为体现个人权利、尊严和财富的一种方式。在古埃及3000多年的历史中，奴隶、农民和其他底层人只穿最简单的衣服，表明了他们的社会阶层微不足道。

Loin cloth was shaped and worn like triangular diapers. It could be the sole garment of lower class men or undergarments for higher class men. Schenti was a wrapped skirt, the length, width and fit of which varied with different time periods and different social classes.

缠腰布的形状和穿法很像一片三角形的尿布。它可以是底层人的唯一服装，也可以是上流阶层的人里面穿的衣服。"申提"是一种缠裹式的裙装，在不同时期、不同社会阶层中其长度、宽度和合体度都有所变化。

Tunic was a T-shaped garment with openings at the top for the head and arms (Figure 3-1). Calasiris (also kalasiris) referred to a loose-fitting robe style garment. Ancient wall paintings showed that it was made of almost sheer linen. Men wore it over loin cloth or a short skirt, or under schenti.

筒形衣是一种T字形的服装，上面有伸出头和胳膊的开口（图3-1）。"卡拉西里斯"是指一种宽松的长袍式服装，从古代壁画上看，它是由一种几乎透明的细亚麻布制成的。男子把它穿在缠腰布或短裙的外面，或者申提的里面。

Sheath was the most common garment for women of all classes. It was closely fitted and appeared as a tube of fabric wrapped around the body from under the breasts to the ankles; the upper part was held in place by one or two straps. Tunic and calasiris were usually seen on women of lower status, rather than on upper class women.

紧身衣是所有阶层女性最常穿的衣服。它很紧身，像用一块管状的布料从胸部以下包到脚踝；上面用一根或两根肩带挂住。筒形衣和卡拉西里斯经常在底层女性身上见到，而上层女子却较少穿用。

Figure 3-1　Tunic, calasiris, loin cloth and sheath
图3-1　筒形衣、卡拉西里斯、缠腰布和紧身衣

3.1.2　Mesopotamian costume 美索不达米亚服装

Mesopotamian costume was most likely made of wool which was described as one of the chief products of Mesopotamia. The variety of fabrics and decorations applied to the costume seemed to be quite complex, so the weaver may have to master a complicated system of manufacture.

美索不达米亚的服装很可能是由羊毛制成的，据说羊毛是美索不达米亚地区的主要产品之一。服装上运用的各种面料和装饰似乎非常烦琐，所以工匠们可能需要掌握一套复杂的制作方法。

Both men and women mainly wore skirts and the type was similar (Figure 3-2). Skirts wrapped around the body. When fabric ends were long enough, the end of the fabric length was passed up, under a belt, and over one shoulder. The skirts were probably first made of sheepskin with the fleece still attached. Even after sheepskin had been supplemented by woven cloth, the cloth was fringed at the hem or constructed to simulate the tufts of wool on the fleece.

男女都以穿裙子为主，类型相似（图3-2）。裙子缠裹在身上；当面料足够长时，将面料向上延伸，束上腰

Figure 3-2　Men's costume of Mesopotamian
图3-2　美索不达米亚男装

带，搭在一只肩膀上。裙子最初可能是由带毛绒的羊皮制作。即使后来羊皮辅以织布，布的下摆也会做一些流苏，或者把织布做成模仿毛绒上羊毛簇的样子。

3.1.3　Ancient Greek costume 古希腊服装

From some wall paintings, we can notice that the Greek costume could be divided into two types: chiton and himation. Chiton was worn as the undergarment, and himation acted as the outerwear. Men and women's costume had little difference between each other, and many styles could be worn by both men and women.

从壁画上看，古希腊的服装可以分为两大类型："希顿"和"希玛申"。希顿穿作内衣，希玛申被用作外衣。男装和女装基本上没有差别，很多款式都是男女同穿的。

Chiton was made of a single rectangular fabric, doubling up and wrapping around the body, securing at the shoulders with one or more pins. The two main types of it were Doric chiton and Ionic chiton (Figure 3-3). Doric chiton was usually made of wool, folded over at the top and fastened with two large straight pins at the shoulders. Ionic chiton was fastened with many small pins and would form two sleeves when belted. It could be made of lightweight wool or linen.

希顿用一块长方形的布料做成，对折后包裹住身体，在肩部用一根或多根别针固定。"多利安式希顿"和"爱奥尼亚式希顿"（图3-3）是其中两种主要类型。多利安式希顿通常用羊毛面料制作，上端翻折，用两根直形别针固定于肩部。爱奥尼亚式希顿用很多小别针固定，扎腰带后会形成两只袖子。它可以用轻质的毛织物或亚麻制作。

Himation was a rectangle of fabric large enough to wrap around the body, and was worn over a chiton. Some philosophers were depicted in the himation alone, without a chiton beneath, but whether this was an artistic convention or actual practice is unclear.

希玛申是一件很大的足以包裹住整个身体的长方形布料，穿于希顿之上。很多哲学家被描绘成只穿希玛申而不穿希顿的形象，但这究竟是艺术惯例还是

Figure 3-3　Doric chiton and Ionic chiton
图3-3　多利安式希顿和爱奥尼亚式希顿

真实的情况却不得而知。

3.1.4　Ancient Roman costume 古罗马服装

The Roman costume was mainly adopted from the Greeks, but there was a most distinctive costume form, toga. Toga (Figure 3-4) was male Roman citizen's costume worn over the tunic. Each variety of togas had distinguishing characteristics in shape, mode of decoration, color, or form of draping. The basic style of toga was draped from a length of white wool fabric, roughly semicircular in shape, with a band of color around the curved edge.

古罗马主要采用了古希腊人的服装，但有件衣服最与众不同，那就是托加。托加（图3-4）是罗马市男性市民穿于筒形衣之上的衣服。每种托加在形状、装饰样式、颜色或垂挂方式上都有所区别。托加的基本款式是由一块白色的羊毛织物披挂而成，这块织物近似椭圆形，弯曲的边缘上还有一条色带。

Figure 3-4　Toga
图3-4　托加

Women of ancient Rome mainly wore stola and palla. Stola (counterpart to the Greek chiton), which was an outer tunic, was worn for out-of-doors and more formal occasions. Palla was a draped shawl (counterpart of the Greek himation) placed over the stola. Sometimes it was pulled over the head like a veil.

古罗马女性主要穿"斯托拉"和"帕拉"。斯托拉（与古希腊希顿对应）是一种外穿的筒形衣，用于户外或较正式场合。帕拉是穿于斯托拉外面的悬垂披肩（与古希腊希玛申对应）。有时会像面巾一样从头上包裹下来。

3.1.5　Costume of the Early Middle Ages (300–1300) 中世纪前期服装（300—1300）

The capital of the Byzantine Empire was Constantinople, situated at the literal crossroads between East and West. Therefore, Byzantine art became a rich amalgam of Eastern and Western cultures. In costume one sees this reflected in a gradual evolution of Roman styles as they added increasingly ornate eastern elements.

拜占庭帝国的首都是君士坦丁堡，位于东西方交界的十字路口上。因此，拜占庭艺术成为东西方文化融合的瑰宝。在服装中，这种融合反映在罗马风格的逐渐演变上，增添了越来越多的华丽的东方元素。

Men's basic garment in the Byzantine period was a tunic. It had various lengths according to the wearers. Some tunics were decorated with clavi on both sides of the front indicating the wearer's status. Such Tunics were named dalmatic (Figure 3-5), on which were decorated some square or round medallions. Women's stola were replaced by dalmatic, and the palla was replaced by a veil which was worn over the head then wrapped around the body and part of the skirt.

拜占庭男装的基本款式是筒形衣。穿者不同，长度不同。有些筒形衣前身的两侧饰有"克拉比"条纹，以显示穿着者的身份。这种筒形衣叫作"达尔玛提克"（图3-5），上面还饰有方形或圆形的徽章。女性的斯托拉被达尔玛提克取代，而帕拉被从头一直包到裙子的面巾取代。

Figure 3-5　Dalmatic
图3-5　达尔玛提克

During the Roman era the primary garments were still tunics for men, and stola and palla for women. After the fall of the Western Roman Empire to A.D. 900, these costumes were blended with those of the Nordic people who moved south. The Nordic people came from colder areas, so their costumes had to be suited to the climate. They wore a type of trousers. Gathered hose which became part of western medieval costume were derived from them.

在西罗马帝国时期，男装主要还是筒形衣，女装是斯托拉和帕拉。西罗马帝国灭亡以后至公元900年，这些服装和移居到南方的日耳曼民族服装融合到一起。日耳曼民族来自寒冷的地区，他们的服装要适应那种气候。他们穿长裤。西方中世纪服装中的"连裆裤袜"便源于他们。

From the 10th to 13th centuries men's and women's costumes were mainly tunics and mantles. The tunic had two types: surcot (outer tunic) and cotte (inside tunic). Men's tunics were a little shorter, while women's tunics were a full-length dress. Women's surcot (Figure 3-6) was usually pulled up and draped over a belt to show the decorative border of the cote.

10世纪到13世纪男女主要的服装是筒形衣和斗篷。筒形衣有两种："苏尔考特"（外穿筒形衣）和"科特"（内穿筒形衣）。男性的筒形衣稍短一点，而女性的筒形衣是一件及地长裙。女性的苏尔考特（图3-6）腰身经常会提起掖在腰带里，以露出里面科特的边缘装饰。

3.1.6 Costume of the Late Middle Ages (1300–1500) 中世纪后期服装（1300—1500）

A number of new garments such as pourpoint, cote-hardie, houppelande came into use (Figure 3-7). Pourpoint (a fitted jacket, also called doublet) was a garment for men, which was worn over the undershirt and tailored to fit the body closely. It was fastened down the front, closing with closely-placed buttons. Strings sewn to the underside of the pourpoint skirt below the waist allowed attaching the hose. Hose were tight bottoms for men.

Figure 3-6   Surcot and cote
图3-6 苏尔考特和科特

一些新的服装如"普尔波万""柯达迪""豪普兰德"开始穿用（图3-7）。普尔波万（紧身上衣，也称为"达布里特"）是一种男士服装，穿于衬衣之上，裁剪得紧身适体。它从前身固定，用排列紧密的纽扣扣住。在普尔波万腰部以下的裙身内侧缝有绳带，能够系缚裤袜。裤袜是男士的紧身下装。

Cote-hardie was a unisex garment tailored to fit the torso. Men's cote-hardie was long to the knee but women's trailed on the floor. The sleeves ended at the elbow, with a dangling lappet falling from behind the elbow. Some women's cote-hardie had a low, round neckline. When wearing it, people would wear a belt at the waist or hip. Sometimes it was also decorated with family heraldries to show the wearer's parentage and social status.

柯达迪是一种裁剪合体，男女皆宜的服装。男性的柯达迪长及膝盖，而女性的则拖到地面。袖长仅到肘部，肘部后面伸出一条悬挂的垂饰。一些女性穿的柯达迪开有很低的圆领口。穿着时，人们会在腰部或臀部配一条腰带。有时它还饰有家徽图案，以显示穿着者的出身和社会地位。

Houppelande was a gown style garment; fitted over the shoulder, then gradually widened below. It had some tubular folds or pleats when held in place by a belt. The style was especially suited to be made of heavy fabrics such as velvet, brocades and wool.

豪普兰德是一件长袍式的服装；肩部合体，然后向下逐渐加宽。系腰带后会出现一些管状的褶皱或褶裥。这种风格非常适合用厚重的面料制作，如天鹅绒、锦缎和毛织物。

Figure 3-7　Pourpoint, cote–hardie, houppelande

图3-7　普尔波万、柯达迪、豪普兰德

3.2 European costume evolution II
欧洲服装演变（二）

3.2.1　Costume of the Renaissance (1400–1600) 文艺复兴时期服装 （1400—1600）

Renaissance literally means re-birth, a return to the civilizations of Greece and Rome. The Renaissance could be viewed as a time of transition from medieval to modern human civilization and view of the world.

"文艺复兴"的字面意思是重生，即再现希腊罗马时期的文明。文艺复兴可以看作是从中世纪跨入现代人类文明与世界观的过渡期。

Men's costumes mainly consisted of a shirt, doublet (a fitting jacket), a jerkin (the etymology of jacket) and hose. Doublet was worn over the shirt and under the jerkin. The modeling of doublet was wide with pads at the front, shoulders and sleeves. The jerkin was similar in shaping as the doublet, but as it usually had short puffed sleeves, under which the doublet sleeves showed out.

男士的服装主要由衬衣、达布里特（紧身上衣）、短上衣（jerkin 为"jacket"的语源）和裤袜组成。达布里特穿在衬衣的外面，短上衣的里面。达布里特造型宽阔，在前身、肩部和袖子上都加有衬垫。短上衣的造型和达布里特类似，但由于袖子通常是短帕夫袖，因此可以露出里面达布里特的袖子。

Ruff was one of the most distinctive details of costume during this period. It was a separated collar which was wide, often made of lace, and had to be supported by a frame or by starching. It could be seen in both men and women's costumes.

拉夫领是这个时期最具特色的服装细节之一。这是一个单独的、宽大的领子，通常用蕾丝制作，而且需要用支架或通过上浆来支撑。在男女装中都能见到这种领子。

Hose were divided into two sections: upper stocks (the upper part, also called breeches or trunk hose) and nether stocks (the lower part). Although upper stocks and nether stocks were sewn together, upper stocks essentially took on the appearance of a separate garment, and were cut somewhat fuller than the lower section.

男子裤袜分为两部分："阿帕·斯道克斯"（上半部分，也称马裤或短罩裤）和"乃惹·斯道克斯"（下半部分）。虽然阿帕·斯道克斯和乃惹·斯道克斯是缝在一起的，但阿帕·斯道克斯本质上具有独立的外观，并且比下半部分裁制得宽肥一些。

Women wore a kind of dress which gave an overall silhouette rather like an hourglass. The dress consisted of a bodice and a skirt, which were sewn together. Bodices narrowed to a small waistline. Women wore a special undergarment named basquine to support their bosom; skirts were supported by a special underskirt called farthingale, which was made of whalebone, cane, or steel hoops. Farthingale mainly had two types: Spanish type and French type. The Spanish farthingale gradually expanded to a cone shape; while the French style expanded abruptly from below the waist. The shape of the former was like a bell, while the latter was like a drum (Figure 3-8).

女士穿一种整体廓型像沙漏的连衣裙，这种裙服是由缝在一起的紧身衣和裙子组成的，紧身衣在腰部收得非常紧窄。女士们穿一种名叫"紧身胸衣"的特殊内衣衬托胸部；裙子是由一种叫作"裙撑"的特殊（内）衬裙支撑的，这种衬裙由鲸须、藤条或钢箍制成。裙撑主要有两种类型：西班牙式和法国式。西班牙式裙撑逐渐扩展成一个圆锥形，而法国式裙撑则是从腰部以下突然扩大。前者的形状像一个吊钟，而后者像一个鼓（图3-8）。

Over the farthingale and under the dress women wore a petticoat. The front of the petticoat was visible through an inverted V shape of the skirt of the dress, while the back of the petticoat was covered. Therefore, only the visible part was made of expensive fabric, while the invisible back was made of less expensive fabric.

在裙撑的外面，裙服的里面会穿一条（外）衬裙。衬裙的前面可以通过裙服下裙身的倒V形开口显露出来，后面被盖住了。因此只有前面可以看见的部分用昂贵的面料制作，而后面不可见的部分用比较便宜的面料制作。

Figure 3-8　Men's and women's costumes of the Renaissance
图3-8　文艺复兴时期的男装和女装

3.2.2　Baroque style costume (1600–1700) 巴洛克风格服装（1600—1700）

The Baroque style emphasized lavish ornamentation, free and flowing lines, and flat and curved forms. These curvilinear forms were also reflected in clothing (Figure 3-9).

巴洛克风格强调奢华的装饰，自由而流畅的线条，平缓而弯曲的造型样式。这些曲线造型也反映在了服装上（图3-9）。

Men's ensemble gradually evolved into the three-piece suit including doublet, a waistcoat and breeches. The doublet was shorter in the earlier period of the 17th century and its skirt-like extension reached to the hip. Lace was decorated on the cuff and rabat (a kind of collar like neckerchief). Later a kind of knee-length dress coat replaced the doublet as the outer garment. Such a coat was fitted at the waist and flared to the knees. It also had fitted sleeves with turned-up cuffs.

Figure 3-9　Men's and women's costumes of Baroque style
图3-9　巴洛克风格的男装、女装

男装逐渐演变为三件套组合，包括达布里特、马甲和马裤。达布里特在17世纪初期比较短，其裙状衣摆长及臀部。蕾丝装饰于袖口和"拉巴"领（一种围巾状的领子）上。后来一种长及膝盖的礼服外套代替了达布里特成为外衣。这种外套腰部贴合，并向膝盖处展开；它还带有翻折袖口的紧身袖子。

Waistcoats were worn under the coats. They were cut along the same lines as outer coats, but slightly shorter and less full; most of them were sleeveless. Breeches were cut either full throughout or more closely and tapering gradually to the knee. The lower edges might be decorated with ribbons or lace.

马甲穿于外套里面。其裁剪和外套差不多，但是要短和瘦一点；多数马甲没有袖子。马裤或做得整体宽松，或做得紧身一些且向膝盖处逐渐变细；马裤的底端可以饰有缎带或蕾丝花边。

The silhouette of women's dresses was a little softer than that during the Renaissance. Lace was often used on the cuffs, neckerchiefs and across the bosoms. Necklines tended to be low; some were V-shaped, some square, and some horizontal in shape.

女装的廓型比文艺复兴时期变得柔和一些。蕾丝经常用在袖口、领巾和胸口上。领口往往开得很低，有V形领、方形领和一字领等。

3.2.3　Rococo style costume (1700–1790) 洛可可风格服装（1700—1790）

Rococo style was marked by "S" and "C" curves, tracery, and scroll-work; it was delicate and vigorous.

洛可可风格主要以"S"形和"C"形曲线、花饰窗格图案和涡旋装饰为主要特征，精致而充满活力。

The major elements of men's costume consisted of dress coats, waistcoats, knee-length breeches and hose (Figure 3-10). Hats and wigs were added on appropriate occasions. This style remained throughout the 18th century, but it had some differences in the first and second halves of the century.

男装主要由礼服外套、马甲、长及膝盖的马裤和长筒袜组成（图3-10），帽子和假发会在合适的场合使用。这种风格贯穿了整个18世纪，但在18世纪前半期和后半期却有所区别。

Figure 3-10　Men's costume of Rococo style

图3-10　洛可可风格男装

In the early 18th century there were more embroidered decorations along the open front and cuffs of the dress coats and on the waistcoats. Stand collars were fashionable. In the later 18th century the frock coat emerged. It was looser and shorter and had a flat, turned-down collar; the embroideries on the front and cuffs were decreased, but the buttons of the cuffs were remained. The front of the waistcoats was still made of elaborate fabrics, while the invisible back was made of plain and inexpensive ones.

18世纪初期在礼服外套的前襟、袖口和马甲上都有大量的刺绣装饰，立领流行。18世纪晚期出现了礼服大衣，它较宽松和短一些，带有平整的翻领；前襟和袖口上的刺绣减少了，但袖口上的纽扣保留下来。马甲的前身还是用精致的面料制作，而看不见的后身则是用朴素廉价的面料制作。

Figure 3-11　Women's costume of Rococo style
图3-11　洛可可风格女装

Women's costumes were mainly gown dresses. The shape of women's costume during the 18th century was provided through a variety of supporting undergarments. The width of the skirt became wider from left to right rather than from front to back, therefore the cross-section of the skirt was like an oval. The front part of the bodice was decorated very beautifully with embroideries, masses of ribbons, artificial flowers or lace (Figure 3-11). Women's waists were tightly bound in corsets which were made of coarse fabrics and were supported by frames at the front and back. If the front part of the corset was visible, it also would be decorated elaborately.

女装主要是袍式长裙，18世纪的女装造型是通过各种塑型内衣塑造出来的。下身裙子的左右宽度大于前后宽度，因此裙子的横截面像一个椭圆形。紧身衣的前面部分用刺绣、大量的缎带、人造花或蕾丝装饰得非常漂亮（图3-11）。女性的腰部又用紧身胸衣紧紧束住，紧身胸衣用粗质的面料制作，前后都有框架支撑。如果胸衣前身显露出来的话，也会被装饰得很精致。

Additionally, there was a looser style described as Watteau back (Figure 3-12). Watteau was an 18th century painter who often depicted women wearing such gowns.

But this term was not used until the 19th century. Watteau back had a fitted front and unbelted back. There were some pleats gathered at the back neck falling down to the train.

另外，还有一种宽松的样式被描述为"华托服"（图3-12）。华托是18世纪的一个画家，经常画穿着这种服装的女士。但是这个名称直到19世纪才开始使用。华托服前身合体，后身宽松。在后领口处聚集着一些褶裥，向下一直拖到裙裾。

At the end of this century a bustle style became fashionable which focused on the hips. The hip pads supplanted the hoop to support the fullness.

Figure 3-12　Watteau back
图3-12　华托服

在18世纪末期一种突出臀部的"巴斯尔风格"成为流行。臀垫代替裙撑来支撑这种丰满的造型。

3.3 European costume evolution Ⅲ 欧洲服装演变（三）

3.3.1 Costume of the Directoire Period and the Empire Period (1790–1820) "执政内阁"与帝政时期服装（1790—1820）

From this period women's clothing was more complicated and more subject to change than that of men. Dresses had a tubular shape, with a waistline placement just under the bosom (Figure 3-13). Raised waistline was a notable characteristic of this period. The fabrics of dresses were supple, thin and lightweight, and sometimes even a little sheer. The sleeves were short, some shapes of which were like a melon. To keep warm, women often wore long gloves. The gloves ended on the upper arm or above the elbow; they were made of leather, silk, or net.

Figure 3-13　Women's costume, 1790–1820
图3-13　1790—1820女装

从这一时期起女装比男装更加复杂和易于变化。裙服是管状的造型，腰线在胸部以下（图3-13）。腰线的提高是这一时期的显著特征。裙服的面料柔软、轻薄，甚至有时还有些透明。袖子很短，有的形状像甜瓜。为了保暖，女士们经常戴长手套。手套的长度长及上臂或超过肘部，由皮革、丝织物或网眼面料制作而成。

The corset and hip pad ceased to be used, replaced by a lightweight chemise. The chemise was made of cotton and linen and worn closest to the body. Drawers became a basic part of women's underclothing. Unlike modern underpants they were usually open through crotch area, an important convenience when women wearing long, bulky skirts needed to relieve themselves.

紧身胸衣和臀垫暂停使用，取而代之的是轻薄的衬裙。衬裙为棉质或麻质，贴身穿着。内裤成为女性内衣的基本组成部分。和现代内裤不同的是它们是开裆的，非常方便穿着长而笨重裙子的女性如厕。

Bonnet was a pretty style hat prevailing till the middle of the 19th. It was made of cloth and straw with artificial flowers or ribbons on the brim.

"包妮特"是一种风格甜美的帽子，一直流行到19世纪中期。它用布料和麦秆制作，边缘饰有人造花或缎带。

The revolution in men's clothing was more subtle, but also more long-lasting, than that in women's dress. While neither the components of men's dress (coat, waistcoat, breeches or trousers), nor the silhouette were radically altered, gone were the colors, the lavish embroideries, and the luxurious fabrics. These changes of men's dress were mainly due to the political factors.

男装的变革比起女装更微妙一些，但也更持久。虽然不管是男装的套装组件（外套、马甲、马裤或长裤），还是廓型都没有发生根本的改变，但是华丽的色彩、奢华的刺绣和奢侈的面料却消失了。男装的这些变化主要是源于政治上的因素。

Men's suits were composed of shirts, waistcoats, coats, breeches or trousers and neckwear (Figure 3-14). The front of the coat ended at the waist, but the back ended slightly above the knee pit. Breeches and trousers were different, breeches ended at the knee, but trousers extended to the ankle. There was an interesting phenomenon in France: trousers were often worn by civilians, while breeches were worn by nobles.

Figure 3-14　Men's costume, 1790–1820
图3-14　1790—1820男装

男士的套装主要由衬衫、马甲、外套，马裤或长裤、领饰组成（图3–14）。外套的前身止于腰部，后身的长度稍高于膝盖窝。马裤和长裤不同，马裤止于膝盖，而长裤及至脚踝。在法国有一个很有趣的现象：平民一般穿着长裤，而贵族穿马裤。

3.3.2　Costume of the Romantic Period (1820–1850) 浪漫主义时期服装（1820—1850）

Romanticism was applied to the literature, music, and graphic arts of the same era. For costume, some romantics wanted to attempt to revive certain elements of historical dress, such as ruffs or large sleeves.

浪漫主义被应用于这个时期的文学作品、音乐和平面艺术中。在服装上，一些浪漫主义者试图想恢复历史上的某些服饰元素，如拉夫领和大袖子。

With the development of industrialization, women's social roles were increasingly confined to the home. The affluent women's daily activities were limited; only the working-class women went out to work because of poverty. Women's costume of this period was less functional, for that dresses had large sleeves and the sleeves were set low on the shoulders. Wearing such dress, women would not be able to raise the arms above their heads and were virtually incapable of performing any physical labor. In the meantime, the skirt of the dress was becoming wider, fitting through the hip, and gradually flaring out to ever greater fullness at the hem (Figure 3-15). A corset was used again and the waistline moved lower than the beginning of this century.

随着工业化的发展，女性的社会角色却越来越限定于家庭中。富裕家庭的女性其日常活动非常有限，只有劳动阶层的女性因为贫穷才出来工作。这一时期的女装机能性不强，裙服的袖子很大，并且袖子在肩部的位置偏低。穿这种服装，女士们不能将手臂伸过头顶，实际上她们不能从事任何体力劳动。与此同时裙服的裙身变得更加宽大，臀部合体，并逐渐向造型饱满的底摆展开（图3–15）。紧身胸衣再次使用，腰线也比19世

Figure 3-15　Women's costume, 1820–1850
图3-15　1820—1850女装

纪初时降低了。

Coat, waistcoat, and trousers or pantaloons were still the components of a man's suit (Figure 3-16). Coats had two styles: the tail coat and frock coat. The tail coat was a dress coat used on formal occasions. It was single-breasted or double breasted. The frock coat was more casual than the tail coat. It was usually worn during the day, while the tail coat was more often used for evening dress. Frock coats were fitting the torso, and at the waist had a skirt flared out all around ending at about knee level. Chest and shoulder pads were also used. In this period, some men even wore corsets to make their waists slender and achieve a fashionable silhouette. Gentlemen would also match a silk hat, a cravat, and a cane.

Figure 3-16　Men's costume, 1820–1850

图3-16　1820—1850男装

外套、马甲、长裤或（长款）马裤仍然是男子套装的组成部分（图3-16）。外套有两种：燕尾服和礼服大衣。燕尾服是礼服外套，用于正式场合，单排扣或双排扣。礼服大衣比燕尾服要休闲一些。它一般用于日间，而燕尾服常用作晚礼服。礼服大衣比较合体，从腰部以下往四周张开，长度及膝。胸垫和垫肩也会使用。这一时期，有些男子甚至会穿紧身胸衣将腰勒细，以塑造时尚的形象。绅士们还会搭配高筒礼帽、领巾和文明杖。

3.3.3　Costume of the Crinoline Period (1850–1869) "克里诺林" 时期服装（1850—1869）

The name Crinoline came from the skirt hoops worn by women. The hoops of this period were wide and stiffened just like a cage. The 18th century fashion style was revived at this time.

"克里诺林"的名字来自女士们穿的裙撑。这时期的裙撑宽大而又僵硬，就像一个笼子，18世纪的服装风格在这个时期复活了。

The sewing machine was invented in this period. Some men's and women's clothing or items could be produced quickly at a lower cost. The first synthetic dye was developed in 1856. This technology increased not only the range of colored fabrics, but also their intensity.

缝纫机在这一时期被发明出来。一些男装女装或服装部件能够高效率、低成本地生产出来。1856年人们又研制了第一种合成染料，这种技术不仅增加了面料的色彩种类，也提高了颜色的纯度。

The basic silhouette of women's costume fitted closely through the bodice to the waist, then the skirt immediately widened into a full round. In the early 1850s, skirts were dome-shaped, and in the 1860s, more pyramid-shaped with fullness toward the back. By the late 1860s the fullness below the waist became lessened.

女装的基本廓型是腰部以上非常合体，腰部以下裙子骤然向外张开，形成一个圆形。1850年代初裙子是圆屋状的；1860年代以金字塔式的居多，并且向后膨胀；到了1860年代末期腰部以下便不那么饱满了。

Women's costume from inside to outside included: a chemise, drawers, a corset, a hoop, a petticoat, a dress, and outdoor garments (Figure 3-17). Hoops were made of either whalebone or steel. Dresses were divided into daytime dresses and evening dresses. Daytime dresses were either one-piece with bodice and skirt seamed together at the waist, princess style without a waistline seam, or two-piece with matching but separate bodices and skirts. Most evening dresses were two-piece having many decorative effects. Skirts were often trimmed with artificial flowers, ribbons, rosettes, or lace.

女装从里向外包括：衬衣、内裤、紧身胸衣、裙撑、衬裙、裙服和户外服装（图3-17）。裙撑用鲸须或钢铁制作。裙服分为日间和晚间两种。日间裙服是衣裙相连的一体式，无腰缝的公主式，或是上衣和裙子搭配的两件套。多数晚间裙服是有很多装饰的两件套。裙子上经常饰有人造花、缎带、玫瑰花结或蕾丝。

Men's suits were made up of three pieces as before (Figure 3-18). The change of men's wearing was subtle and unnoticeable. Tail coats were made with velvet-faced lapels. Some trousers were wider at the top and narrowing gradually to the ankles. Trousers and waistcoats often had buckles which were used to adjust their fit. Shirts also had daytime and evening styles. Daytime shirts had less decoration, while evening shirts had embroidered or ruffled fronts.

男装和以前一样还是三件式套装组合（图3-18）。男装的变化微妙且含蓄。燕尾

Figure 3-17　Women's costume, 1850–1869
图3-17　1850—1869女装

服带有天鹅绒面的翻领。一些裤子的上面比较肥，然后逐渐向踝部变细。裤子和马甲经常带有可以调节松紧的带扣。衬衫也分为日间和晚间两种。日间衬衫装饰较少，而晚间衬衫会带有绣花或褶皱的胸饰。

3.3.4 Costume of the end of 19th century (1870– 1900) 19世纪末期服装（1870—1900）

There were two styles of women's costume in this period: Bustle style (1870-1890) and "S-shape" style (1890-1900). The Bustle Period derives its name from the bustle, a device which provided the shaping for a silhouette with marked back fullness. The fullness was supported by the hoop or pad, or by concentrating many heavy draperies at the back. The skirt was bulky, but the bodice was still tight (Figure 3-19).

Figure 3-18   Men's costume, 1850–1869

图3-18   1850—1869男装

这一时期的女装有两种风格：巴斯尔式（1870—1890）和 "S形" 式（1890—1900）。"巴斯尔时代"的名称来自巴斯尔，是一种让后身造型丰满的装置，这种饱满的造型由裙撑或臀垫支撑，或由在后身堆积的厚重垂褶产生。裙子的体积庞大，但上半身仍然紧瘦（图3–19）。

In the 1890s the bustles were not so exaggerated. Sleeve styles were large and wide at the top; waists were as small as corsets could make them; and the skirt flared out into a bell shape. From the side, the body line resembled an S-shape.

到了19世纪90年代巴斯尔造型不再那么夸张了。袖子的上部变得宽大，腰部尽可能用紧身胸衣勒细，裙子向外张开呈一个钟形。从侧面看上去，身体线条好似一个S形。

A sack suit jacket for men's evening dress was introduced from the United States, which was known as Tuxedo. It was a simplified tailcoat and more casual. Shirts had stiff, starched shirt fronts for formal daytime wear; and pleated shirt fronts for evening dresses. Standing stiff collars of shirts were wide and bow ties were popular.

男式的晚礼服从美国引进了一种宽松式的西服，名叫"塔士多"。它是一种更加休闲的简化式燕尾服。带有硬挺、浆制胸饰的衬衫用于日间正装，带有褶裥胸饰的衬衫用于晚装。衬衫立领高且笔挺，并流行戴蝴蝶结。

FASHION DESIGN ENGLISH 服装设计英语

巴斯尔式　　　　　　　　"S形" 式　　　　　　　　塔士多

Figure 3-19　Women's and men's costumes, 1870–1900
图3-19　1870—1900男女服装

3.4 European costume evolution IV
欧洲服装演变（四）

3.4.1　1900–1918 style 1900—1918年风格

Women's costume style had no radical changes from the 1890s to 1914. The silhouette also could be described as an S-shape. One-piece dresses were more popular. Skirts were flat in front and emphasized a rounded hipline in the back. After hugging the hips, skirts flared out to a trumpet shape at the bottom.

从1890年代到1914年，女装风格并没有发生明显的变化，其廓型还可以被形容为S形，一体式的连衣裙比较流行。裙子的前身比较顺直，后身突出圆润的臀部；裙子上半部分紧包住臀部，向下呈喇叭花状展开。

World War Ⅰ lasted from 1914 to 1918, and influenced costume style greatly. The most obvious influence was that women's costumes became more comfortable and practical. These clothes were required for women's more active participation in the variety of jobs that they had taken over from men. Military elements were evidently used in the cut of women's clothing. Skirts grew short and narrowed; S-shaped curves were being superseded by straighter lines. The fit of dresses waist was looser and being belted at the waist were popular.

第一次世界大战从1914年持续到1918年，它对服装风格的影响很大。最明显的影响就是女装变得舒适而实用，这些服装是女性在更积极地参与到从男性手中接管的各项工作中时所需要的，军装元素也在女装的设计裁剪中体现明显。裙子变短变窄，S形曲线正被直线形取代，裙服的腰身变得宽松并流行在腰部扎腰带。

Suits, consisting of jacket, waistcoat, and trousers and worn with a shirt and necktie were appropriate dress for professional and business men during the work week. Suits were worn for social occasions by blue-and white-collar workers, as well. For informal social occasions or during their leisure time, men could wear sport jackets, trousers, and shirts of various kinds. After the war, some of the soldier clothing passed into use by the general public, such as trench coat. It was a water-repellent coat with a belt at the waist, and then became a standard item of rainwear for men.

由上衣、马甲和长裤组成的西服套装，搭配衬衫和领带是适合职业和商务男士在工作时穿着的装束；蓝领和白领工人在社交场合都穿西服套装。在非正式社交场合或业余时间，男士们可以穿运动上衣、长裤和各式衬衫。第一次世界大战以后，有些士兵服被一般大众穿用，如"战壕式风衣"，它是一种系有腰带的防水外套，后来成为男士们标准的防雨衣。

3.4.2　1920–1947 style 1920—1947年风格

For women in the 1920s, a figure with a flat bosom and no hips was the ideal. The fashionable silhouette was straight, without indentation at the waistline. When a dress had a belt, the belt was placed at hipline. In addition to one-piece dresses, tailored suits which had matching jackets and skirts were also popular; the jacket ended at the hip or below. In the 1920s to 1930s, the skirt hemlines began to lengthen and belts moved gradually closer to the natural waistline. The silhouette of the 1930s emphasized the natural form of the women's body. Bosom, waistline, and hips were clearly defined by the shape of clothing.

对20世纪20年代的女性来说，理想的体型是平胸、扁臀。流行的服装廓型是不显腰身的直线形；裙子外面若是扎腰带的话，也是扎在臀围线上。除了一体式的连衣裙，上衣和裙子搭配的西服套装也很流行；上衣的长度到臀部或以下。20世纪20—30年代，裙子的底摆开始加长，腰带逐渐移至自然腰节处。20世纪30年代的服装轮廓强调自然的女性身体形态，胸、腰、臀可以通过服装造型明显地区分开来。

In the 1940s, influenced by the World War Ⅱ, skirts had become shorter, ending just below the knee. Shoulders of jackets had broadened, and shoulder pads were inserted into garments to provide greater width. Bias cut was rarely used. Throughout the period from 1920 to 1947, women were becoming more active in participating sports, so many women sportswear were designed.

20世纪40年代受第二次世界大战的影响，裙长缩短了，长度刚到膝盖以下。上衣的肩部变宽了，并加入垫肩以增加宽度。斜裁工艺很少使用了。从1920—1947年，女性越来越积极地参与到体育运动中，很多女式运动装也被设计出来。

No dramatic changes in men's clothing took place in this period. Jackets, vests (waistcoats), and trousers of men's suits matched in color and fabric (Figure 3-20). Sack suits were also popular. Men's coat styles included chesterfields and raglan-sleeved coats, with either buttoned front closings or fly-front closings. During this period a whole new category of clothing for men developed. It is generally classified as sportswear, but would perhaps be more accurately termed "leisure" or "casual" clothing. It was worn not only for active sports but during a man's leisure time.

这一时期的男装并没有发生巨大的变化。套装中的西服、马甲和长裤在颜色和面料上互相匹配（图3-20）；休闲西服也很流行；外套包括"柴斯特菲尔德外套"式和带门襟或暗襟的插肩袖式。这时一种全新的男装种类应运而生，它通常被归为运动装类，但或许称之为"休闲装"或"便装"更加准确，它不仅用于体育运动，也用于男士们的闲暇时光。

Figure 3-20　Women's clothing, 1920s; Women's clothing, 1930s; Women's clothing, 1940s; Men's clothing, 1940s

图3-20　20世纪20年代女装、20世纪30年代女装、20世纪40年代女装、20世纪40年代男装

3.4.3　1950s style 20世纪50年代风格

Seldom does fashion change almost overnight, but in 1947 an exceptionally rapid shift in styles took place. The shift referred to a new women clothing style which was different from the wartime period. It was known as the New Look (Figure 3-21) and designed in 1947 by Christian Dior, a French designer. It caused a sensation and was accepted rapidly, then became the basis of style lines for the next decade. Since then the elegant A-line silhouette was almost dominated the whole 1950s' fashion style (Figure 3-22) and lent women's clothing mature femininity charm.

时尚很少在一夜之间发生改变，但1947年一场不同寻常的剧变却在时尚界发生了。这场剧变是指与战时不同的一种全新的女装风格，它被称作"新样式"（图3-21），是由法国设计师克里斯汀·迪奥在1947年设计的。它曾轰动一时，且迅速地被人们接受，并成为而后十年女装造型的基础。自那以后，优雅的A字廓型几乎主导了整个50年代的时尚风格（图3-22），赋予了女装成熟的女性魅力。

Although the style of men's wear did not change radically compared to the past, over time, the clothing styles for men became more diverse. The diversity mainly reflected in the details.

尽管男士的穿着和以前相比没有明显的变化，但随着时间的推移，男装的样式也变得越来越多样化，这些多样化主要体现在细节上。

Figure 3-21　New look
图3-21　"新样式"

Figure 3-22　Women's clothing, 1950s
图3-22　20世纪50年代女装

3.4.4　1960s style 20世纪60年代风格

In this period there was an important group of people who influenced fashion considerably. This group was mainly young people, who were confident, energetic, having different ideals of society, lifestyles and fashion. This group was known as the Beatniks in the United States. These young people intended to use their clothing to make a statement about their differences, but curiously these counter-cultural fashions provided nourishment for the fashion industry, ever-hungry for new ideas.

这个时期有一群重要的人对时尚产生了巨大的影响。这个群体主要是年轻人，他们自信而充满活力，对社会、生活方式和时尚有着不同的憧憬。在美国他们被称为"垮了的一代"，这些年轻人意图通过服装来表达他们的与众不同，但奇怪的是，这些非主流文化的穿着时尚，却给一直如饥似渴地寻找新思路的服装产业提供了养分。

Jeans originated from the American West in the 1850s and initially worn as jumpers by workers, but in the 1960s it was used by common people including both men and women (Figure 3-23). Young people began to use jeans as a medium of self-expression, and before long it became an international hot fashion item.

牛仔装是19世纪50年代在美国西部出现的，最初是用作工人的工作服，但在20世纪60年代普通大众无论男女都在穿用（图3-23）。年轻人用它来表现自我，而且没过多久它便成为国际时尚的宠儿。

Figure 3-23　Jeans, 1960s
图3-23　20世纪60年代的牛仔装

Skirts were becoming shorter and shorter during this period. Miniskirt and Micro mini appeared and they were fanatically pursued by young girls. During the past women were forbidden from wearing pants to work or to school, but during this period pants and pantsuits gained acceptability not only for leisure, but also for all occasions.

这个时期裙子变得越来越短，迷你裙和超短裙出现了，并受到年轻女孩的热烈追捧。过去是禁止女性在工作或学校场合穿裤装的，但在这个时期不但在休闲场合，而且在所有的场合女性穿裤子或裤套装都得到了认可。

Men wearing sportswear was more and more prevalent. Slacks, sports jackets, sport shirts and sweaters were all of this type of clothes.

男士们穿休闲装越来越普遍。宽松裤、运动外套、运动衫和针织衫都是这种类型的衣服。

3.4.5　1970s–1990s style 20世纪70—90年代风格

As a great deal of women entered the work force in the 1970s, the fashion changed. More women entered managerial positions in the corporate world, a so-called power suit was being recommended for women who wanted to "dress for success". This suit was a feminized version of the men's business suit, consisting of a tailored jacket, a moderate-length skirt and a tailored blouse. Some other unisex clothing also emerged, such as blue jeans, T-shirts, sweaters and so on.

20世纪70年代随着大量女性进入职场，时尚也发生了变化。更多女性在企业界加入了管理岗位，一种所谓的"权力套装"正在被推荐给那些想要"穿出成功"的女性。这是一种男士商务套装的女性化版本，由西服、中长裙和精制衬衫组成。其他还出现了一些中性化的服装，如蓝色牛仔装、T恤、针织衫等。

The prevalent silhouette in the 1970s was described by fashion writers as "an easier and more casual fit". The use of softer fabrics that molded the body, displayed body curves. In the 1980s, an emphasis on the shoulders had appeared and shoulder pads were commonly used in women's suits that made them look stronger. Skirts were narrow so the overall silhouette of the suits was like a T-shape (Figure 3-24). After the 1990s, along with the women's clothing variety increased, women never had such a large freedom of choice in wearing.

20世纪70年代流行的服装廓型被时尚作家描述为"更轻松、更随意的搭配"。运用塑造体型的柔软面料，来展示人体的曲线。到了20世纪80年代，出现了强调肩部的廓型，肩垫常常用于女性套装中，让她们看起来更强壮（图3-24）。裙身很窄，所以套装的整体廓型像一个T形。90年代以后，随着女装种类的增加，女性在穿着上从来没有过这么大的选择自由度。

Men's clothing was undergoing the so-called Peacock Revolution since the 1970s. Men, like women, had a great variety of clothing types available to choose. Although the corporate world still required traditional business attire, there was wide latitude in acceptable dress for social occasions and leisure activities. Fashion segmentation increased, and the differences between clothing for

Figure 3-24　Women's clothing, 1970s
图3-24　20世纪70年代女装

active sports and sportswear had been lessening. This change came from men's more casual lifestyle.

自20世纪70年代以来，男装正经历着一场"孔雀革命"，和女士们一样，男士也要拥有大量的服装类型可供选择。虽然在企业界仍然需要穿着传统的商务装，但在社交场合和休闲活动中，服装穿着却有着很大的选择空间。时装被细分化，且运动装和休闲装之间的差距正在缩小。这种变化源自男士们更随意的生活方式。

A London street fashion Punk Style became a factor in influencing some mainstream fashions in the 1970s (Figure 3-25). Some young people wore messy, baggy or torn clothes. Boys generally wore black leather with rivets; girls wore micro-minis with black fishnet stockings and black eye makeup. Moreover, they often dyed their hair brown, yellow or red.

20世纪70年代一种伦敦的街头时尚"朋克装"成为影响主流样式的一个因素（图3-25），有些年轻人穿着凌乱、松垮或撕裂的服装。男孩们通常穿着带有铆钉的黑色皮革服装；女孩们穿极短的超短裙配黑色的渔网袜，并涂抹黑色的眼影；他们还经常把头发染成棕色、黄色或红色。

Figure 3-25 Punk style clothing
图3-25 朋克风格服装

本章小结

■ 服装史的发展与同时期的历史文化背景有关，服装是这些背景的折射与反映。

■ 男装女装都经历了一个由简趋繁，又由繁趋简的过程，男女装有不同的功能诉求。

■ 同一种服装在不同时期有不同名称，或同一名称的服装有不同的表现，具体视历史情况而定。

课后练习

1. Please retell the fashion style and primary costume characteristics of each period.

请复述每个时期的时尚风格和主要的服装特征。

2. Group exercise 小组练习

Please work in groups of three or four, and discuss and exchange your opinions about the costume of some periods.

三或四人一个小组，请对某些时期的服装进行讨论和交流。

Chapter 4
第四章

Fashion Styling
服装造型设计

课题名称： Fashion Styling

课题内容： 服装造型设计

课题时间： 2课时

教学目的： 让学生学习并了解服装造型的原理、服装造型要素、廓型的特点和服装造型细节设计的基本英语表达。

教学方式： 结合PPT与音频等多媒体教学课件，以教师课堂讲解为主，学生随堂练习与小组讨论为辅。

教学要求： 1. 掌握相关词汇。

2. 能够用英语描述出某种服装造型的款式特点和设计细节。

课前（后）准备： 课前预习服装造型基本原理、造型要素、服装廓型和细节设计的相关内容，浏览课文；课后记忆单词，并复述文中内容。

4.1 Basic principles of fashion styling
服装造型原理

4.1.1　Proportion 比例

Proportion refers to the linear sub-division of objects and shapes, and concerns the balance of shape, volume, color, fabric, texture and scale. The combinations of these elements make fashion design infinitely diverse.

比例指的是物体和形状的线性再分，涉及形状、体积、色彩、面料、材质和大小的平衡关系。把这些元素组合在一起可以使服装设计变化无穷。

Having a "sense of proportion" can be said to be subjective, in that the divisions in shapes can appear to be "right" or "wrong" depending on personal view and contemporary values. However, there is a famous rule – the "golden mean" can be used as reference. The idea of the "golden mean" was developed by taking actual measurements of ancient sculpture where it was found that a 5:8 proportional relationship usually existed between the sections of which these figures were composed. There is a general consensus that these proportions are also pleasing when applied to other areas of art, including fashion design. See figure 4-1, apparently the left model looks taller.

"比例感"的出现可以说是非常主观化的，因为依据个人的观点和时代价值观的不同，对图形分割的看法可能是"对"的，也可能"不对"。不过有一个著名的法则——"黄金分割"可以用作参考。"黄金分割"的理念是通过对古代雕塑的实际测量得出的，在那里人们发现，塑像的组成部分之间通常存有5∶8的比例关系。人们普遍认为，这种比例关系应用于其他艺术领域也能取得满意的效果，包括服装设计。如图4-1所示，显然左边的模特看起来更高挑一些。

However, these classical proportions are not always fashionable; "out of proportion" has been equally popular. Fashion flips between the orthodox and traditional to the alternative and challenging; because of this

Figure 4-1　The proportion of 5:8 versus 1:1
图4-1　5:8对比1:1的比例

the golden mean has not been adopted as an absolute rule.

　　然而，这种经典的比例关系并不是一成不变的，"打破比例"同样很受青睐。时尚就是在正统和传统，与另类和挑战之间来回徘徊的；正因如此，黄金分割并没有被用作绝对的定理。

4.1.2　Symmetry and Asymmetry 对称和非对称

　　Symmetry is absolute balance. Bilateral elements can overlap each other. The scale tips neither one way nor the other. The characteristics of symmetry include: the correspondence in size, form, and arrangement of parts on opposite sides of a point, line, or plane; regularity of form or arrangement in terms of like, reciprocal, or corresponding parts. Symmetry need not mean boring; actually, it opens up room for a lot of improvisations.

　　对称就是绝对的平衡，轴线两侧元素可以相互重叠。重心不倾向于任何一边。对称的特点包括：以点、线、面为中心两侧部分的大小、形状和排列一致；在相同、相对或相应的部分上，形状或排列规则整齐。对称并非是枯燥乏味的，实际上，它为很多即兴创作提供了空间。

　　Asymmetry is lack of balance or symmetry. Asymmetry can be achieved in three ways: by equal volume and unequal impact, by unequal volume and equal impact or by unequal volume and unequal impact. The more elements that are contained in the design, the more complex it is to create a pleasing, balanced end result.

　　缺乏平衡或对称感就是非对称。非对称可以通过三种方式获得：量感相同力度不同，量感不同力度相同，或量感和力度都不同。设计中包含的元素越多，要创造出赏心悦目、均衡的效果就变得越复杂。

4.1.3　Balance 均衡

　　A garment achieves balance from the relative volume and size of the style lines and details used in its design. This sense of balance is based on a complex range of comparisons; it is very subjective and also relates to contemporary aesthetic criterions. It can be achieved from symmetric or asymmetric elements. For an asymmetric design, it needs some details to be added at some point to achieve balance. The design only becomes successful when it satisfies the eye in terms of its balance.

　　服装中的均衡，来自设计中运用的款式线条以及细节的相对量感和尺寸。均衡感是建立在一系列复杂对比的基础上；它很主观化，并且和当代的审美标准有关。它可以从对称或非对称的元素中产生。对于非对称的设计，需要在某些情况下添加些细节来取得均衡。只有从均衡的角度看起来比较舒服时，设计才会成功。

4.1.4　Contrast 对比

Using positive and negative effects in color blocking is a useful design development exercise where dramatic differences in balance and proportion can be achieved, as figure 4-2 shows. A positive creates a negative and vice versa, these elements create rhythm when repeated. An optical "dazzling" effect can be created when the two elements are used in equal measure or too closely together. Positive and negative act as a separator, creating maximum visual impact, and can be used not only for color contrast but for stiff and soft, coarse and smooth, solid and void, pattern and plain. The designer can use this powerful tool in order to focus attention on or away from a feature of the design.

在对比色中运用阴阳对比效果是一项有益的设计拓展练习，从中可以获得均衡和比例上的明显差异感，如图4-2所示。阴阳互生，这两种对立元素交替运用便会产生节奏。当阴阳元素的面积对等或距离较近时，就会产生一种视学上的"眩目"效果。阴与阳充当着分隔符的作用，可以产生最大的视觉冲击力，不仅可以用于色彩对比，还可以用于硬挺与柔软、粗糙与光滑、实与虚、花色与素色的效果对比中。设计师可以运用这一有力的工具，来突出或淡化设计中的某一特征。

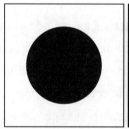

Figure 4-2　Contrast
图4-2　对比

4.1.5　Rhythm 节奏

The use of rhythm as a design tool is important in achieving pleasing design effects. Rhythm results from design elements that form repetitions; these repetitions can be either uniform or of decreasing or increasing size. A regular repeated or changed pattern can generate the sense of rhythm.

作为一种设计工具，节奏的运用对于取得满意的设计效果很重要。节奏源自设计元素的重复，这些重复既可以是均匀一致的，也可以是尺寸的递减或递增。有规律的重复或变化模式可以产生节奏感。

4.2 The elements of fashion styling
服装造型要素

Point, line, plane and body are the four elements of fashion styling. When given a certain form, color and texture, they can make fashion design produce infinite forms.

点、线、面、体是服装造型中的四大要素。当被赋予一定的形、色和质时，它们就可以让服装设计创造出无限的形式。

4.2.1　Point 点

In terms of geometry, point does not have size but only position. In fashion design, the point can be a button, a purse, a brooch or a shape on a garment (Figure 4-3). It can be round, square, floral or irregularly shaped, and also can be planar or three-dimensional. Point is easy to be a visual focus and draw people's attention, so it plays a role in accentuation and embellishment.

在几何学上，"点"没有大小只有位置。在服装设计中，点可以是一颗纽扣、一个手袋、一个胸针或者是服装上的一个造型（图4-3）。它可以是圆形、方形、花形或不规则形状的，也可以是平面或立体的。点容易形成视觉中心，吸引人们注意，因此它起到强调和点缀的作用。

Figure 4-3　Points on the garment
图4-3　服装上的"点"

4.2.2　Line 线

The line of the garment generally refers to its cut, where seams and darts are placed on the garment and the effect that they have visually. Some design elements which take on linear forms are also regarded as lines, such as belts, ribbons, zipper tapes, or necklines. Furthermore, some designers will refer to the line of a garment when they actually mean its silhouette. Line is an important segmentation element in fashion styling. It is the result of edges meeting, marking separations. Some garment lines are concrete, like seams, stitches, and darts; some are abstract, which actually do not exist on garments, but we can recognize visually, like pleats and folds (Figure 4-4).

服装中的"线"通常是指它的裁剪，即接缝和省道安排在服装上的位置和它们具有的视觉效果；一些具有线条形式的设计元素也被看作是线，像腰带、飘带、拉链或领线等。有些设计师在谈及服装廓型的时候，常常会提到线。线是服装造型中重要的分割元素，它是边缘相遇的结果，标志着被分割部分的彼此分离。有些服装的线条是具象的，像接缝、明线和省道；有些线条则是抽象的，它其实并不存在于衣服上，但我们可以通过视觉辨认出来，如褶裥和褶皱（图4-4）。

Figure 4-4　Lines on the garment ("Chuhetingxiang"Spring/Summer 2016 Fashion Show)
图4-4　服装上的"线"（2016春夏"楚和听香"时装发布）

4.2.3　Plane 面

Plane is two-dimensional and occupies a particular area of space. In fashion design the plane has a certain shape; it can be some part of a garment, such as the front panel, the back panel, the lapel, or the pocket, and also can be some patterns on the garment. Planes of different sizes have different visual impact. Large plane has stronger visual impact; the opposite, weaker. Plane can be divided into regular and irregular ones. Generally regular planes have a specific shape, such as rectangle, triangle, circle, or oval and so on. The shape of irregular planes is random and depends on the designer's ideas, as figure 4-5 shows. The design effect of regular planes is simple and rhythmical, while of irregular planes is vivid and full of fun.

"面"是二维的，并占据一个特定的区域。在服装设计中，面具有一定的形状；它可以是服装上的某个部分，如前片、后片、翻领或口袋，也可以是服装上的图案。不同大小的面具有不同的视觉冲击力。大的面视觉冲击力较强；反之，较弱。面可以分为规则形和不规则形。一般规则的面有具体的形状，如长方形、三角形、圆形或椭圆形等；不规则的面形状是随机的，它取决于设计师的创意，如图4-5所示。规则的面设计效果简洁而富有节奏感；不规则的面其设计效果生动而有趣味。

Figure 4-5　Planes on the garment
图4-5　服装上的"面"（2017秋冬龚航宇成衣发布）

4.2.4 Body 体

Body has a three-dimensional property and occupies a certain amount of space. In styling design, it has the strongest visual impact. For fashion design, it can be divided into basic styling body and decorative styling body. For the former, as the human body itself is three-dimensional, designers create shapes to encase, echo, or imitate it. For the latter, the decorative styling body is added to the body's original form; it changes the appearance of the human shape and builds a new silhouette, as figure 4-6 shows. Therefore, it may be exaggerated, creative or even weird; the materials to create this styling can be unusual, like metal, paper, plastic, or even balloons.

Figure 4-6　Body on the garment
图4-6　服装上的"体"

"体"具有三维的性质，占据一定的空间。在造型设计中，它具有最强的视觉冲击力。对服装设计来说，它可以分为基础造型的体和装饰造型的体。对于前者，因为人体本身就是立体的，设计师创作的造型用以容纳、回应或模仿人体。对于后者，装饰造型体是附加在人体原始形态之上，它改变了人体形态的外观，塑造了全新的廓型，如图4-6所示。所以，体可能是夸张的、创意性的，甚至是怪异的；塑造这种造型的材料也可以是不同寻常的，如金属、纸张、塑料，甚至是气球。

4.3 Silhouette
服装廓型

Our first impression of an outfit when it emerges on the catwalk is created by its silhouette, which means that we look at its overall shape before we interrogate the detail, fabric or texture of the garment. Silhouette plays a big part in the initial design. Color, pattern, and detail fill the body of the garment, but the silhouette provides a structural setting for the design concept. Silhouette is a fundamental consideration in fashion design decision making, because it will influence people's perception of an outfit's existence in the surroundings.

当一套衣服在T台上出现时，带给我们的第一印象便来自它的廓型，也就是说在我们审视一件衣服的细节、面料或质地以前先看它整体的形状。廓型在设计之初具有重要的作用。如果说颜色、图案和细节填充了服装的主体，廓型则提供了设计理念中的结构设置。廓型是制定服装设计方案时一个基本的考虑因素，因为它会影响到人们对服装在周边环境中存在的感知。

We can create the silhouette by way of combining geometrical shapes. For example, a triangular shape can create an A-line or trapeze frame; this silhouette flares out to skim over the body. Inverting the triangle so that it tapers toward the bottom can be described as a V frame. When this shape is paired with a vertical rectangle, it is defined as a Y silhouette. Flipping the Y creates an attractive trumpet-like shape. The hourglass silhouette consists of two inverse triangles. Round and oval shapes are effective in adding mass where desired. Garments are transformed by silhouette and the ratio of combinations. A fitted bodice paired with a pegged skirt will have an hourglass appearance, while a full shirt or blouse tucked into a straight pant will give the impression of a Y. A dress with padded shoulders and a tapered hemline results in a T shape. A coat which is loose at the waist but tight at the bottom hem will produce an O shape.

我们可以用组合几何图形的方式来创建廓型。例如，三角形可以创建A型或梯型框架，这种廓型向外展开掠过人体；把三角形倒过来，它逐渐向底端变窄，可以形容为V型框架；当把V型和竖直的长方形搭配时可定义为Y型；翻转Y型便可创建出一个优美的喇叭型；沙漏型是由两个相反的三角形组成的；圆形和椭圆形在需要的时候可以有效地增加体积。服装就是通过廓型和其组合的比例进行变换的。紧身上衣和楔形裙搭配会有沙漏似的外观；而把宽松的衬衫塞进直筒裤里会给人一种Y型的印象；衬有垫肩但下摆逐渐变细的连衣裙会呈现T型；腰部宽松但底端收紧的外套会形成一个O型。

As a collection is presented, a repeated contour of the figure is used in a similar manner in order to communicate the collection's emphasis. Actually, the interpretation of this silhouette will vary in shape, fabric weight, placement on the figure, and proportion, but having a defined contour feature gives the collection a sense of cohesion (Figure 4-7). Finally, note that a fashion silhouette is not defined by the garment alone—the hairstyle, the figure, shoes, mix of garment pieces, and accessories all contribute greatly to the final shape and effect; thus, the thing becomes more complex and new looks are created as shapes play off one another.

当一个服装系列发布时，重复的形象轮廓会以相似的方式被运用，来传达这个系列的重点。实际上，对廓型的诠释会因造型、面料重量、在人体上的设置和比例关系而有所差异，但有一个清晰的轮廓特征，会使整个系列有一种凝聚力（图4-7）。最后，需要注意的是，服装廓型并不仅由服装本身决定——发型、体型、鞋品、服装部件的组合和配饰都会对最终的造型和效果起到很大的作用；廓型的塑造是一项复杂的工作，新的服装外观就在各种造型的相互博弈中被创建出来。

Figure 4-7　Silhouette design (Designer：Lin Lin)
图4-7　服装廓型设计（设计：林琳）

4.4 Details of fashion styling
服装造型细节

An outfit can have a dramatic silhouette and good line, but without great detailing it would appear amateurish and incomplete. Outfits that lack detail can survive on the catwalk, but will not bear close scrutiny, especially when displayed on the mannequin in a shop. Details in clothes are often the clincher when it comes to persuading someone to part with their money. Detailing is particularly important in menswear design, as menswear is not changed as dramatically as women's wear.

服装可以有引人入胜的廓型和美观的外表，但如果没有精致的细节，就会显得设计不专业和不完善。缺乏细节的服装可以在T台上展示，但经不起推敲，尤其是在商店里模特架上展示的时候，服装细节往往是说服某人花钱购买的关键因素。细节在男装设计中尤为重要，因为男装并不像女装那样变化显著。

Details are practical considerations: which fastening to choose, which type of pocket to have and how much top-stitching to use. Clever use of detail can also be used to give a collecting of clothes a unique identity or signature; cutting a pocket in a particular way, using an embellishment in one area of a garment or the finishing of an edge can help to differentiate the garments of one designer from those of another (Figure 4-8).

细节是很实际的考虑因素：如选择哪种固缝方式，运用哪种类型的口袋，用多少明线等。巧妙的细节运用还可以赋予服装系列一种独特的辨识度和鲜明特征，如用特殊的方式裁剪口袋，修饰服装的某个部位，或对服装边缘进行后处理，都有助于把某位设计师的服装和另一位的区别开来（图4-8）。

Thinking about pocket types and fastenings might seem a little mundane, but whether a design is successful or not generally depends upon such details. Although a pocket, for example, has a generic function and the concept is fixed, it does not mean that it needs to be conceived in a formulaic way. There are rules about how certain pockets are made and how these should look, but these notions can and should be distorted and reinvented. Design rules are made to be broken, after all.

虽然考虑运用何种类型的口袋和固缝方式看起来似乎平淡无常，但设计的成功与否往往就取决于这些细节。举个例子，口袋虽然具有通用的功能，其概念是固定的，但这并不意味着它的设计就需要按照程式化的方式处理。尽管制作口袋和设计口袋造型有一套规律，但这些观念可以也应该被转变和彻底颠覆。毕竟，设计规则就是用来打破的。

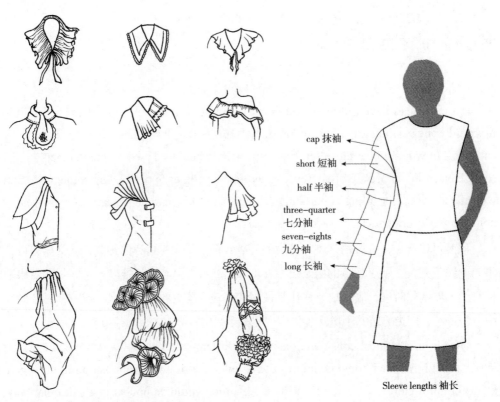

Sleeve lengths 袖长

Figure 4-8　Details of fashion styling (Designer: Pang Bo)
图4-8　服装造型细节（设计：庞博）

本章小结

■ 服装造型的基本原理有比例、对称和非对称、均衡、对比和节奏等。

■ 点、线、面、体是服装造型的四大要素。

■ 服装廓型主要有A、X、Y、T、O型等。

■ 细节也是决定服装造型设计成败的重要因素。

Exercises 课后练习

1. Please search for some fashion show pictures and analyze the principles of fashion styling they applied.

请搜寻一些时装发布会的图片，分析一下所运用的服装造型原理。

2. As the first exercise, analyze the styling elements of these garments in the pictures.

像第一个练习一样，分析一下这些图片中服装的造型元素。

3. Analyze the silhouette and design details of these garments as well.

同样地，分析一下这些服装的廓型和设计细节。

Chapter 5
第五章

Colors and Patterns for Fashion Design
服装设计色彩与图案

课题名称： Colors and Patterns for Fashion Design

课题内容： 色彩理论与色彩调和服饰图案

课题时间： 2课时

教学目的： 让学生学习并掌握色彩的基本理论、色彩文化的叙述与表达，服装图案设计与运用的叙述和表达。

教学方式： 结合PPT与音频等多媒体教学课件，以教师课堂讲解为主，学生随堂练习与小组讨论为辅。

教学要求： 1. 掌握相关词汇。

2. 能够用英语描述出服装色彩和图案的设计风格与运用特征。

课前（后）准备： 课前预习关于服装色彩和图案的基础理论知识，浏览课文；课后记忆单词，并复述文中内容。

5.1 Color theory and color harmony
色彩理论与色彩调和

5.1.1 Color theory 色彩原理

Although the way in which a designer chooses to adopt colors is generally an issue of personal taste, it is necessary to know some color theory which can help the designer create an appropriate palette.

虽然挑选运用色彩的方式通常是个人品位的问题，但了解一些色彩原理很有必要，它可以帮助设计师做出恰当的配色方案。

Hue 色相
It refers to the difference between one color and another, that is, the original characteristic of a specific color, such as red, yellow, green, and blue (see the color page). A color can be identified by its hue.

色相指颜色与颜色之间的区别，也就是一种颜色本来的特征，如红色、黄色、绿色和蓝色（见彩页）。色相可以用来识别颜色。

Value 明度
This refers to the lightness or darkness of a color. Colors mixed with pure white are tints; and mixed with pure black are shades. Different amounts of white or black added to the color will give different shaded results. If colors are arranged according to their brightness or darkness, they will present a rhythm of gradation (Figure 5-1).

明度指颜色的明暗程度。颜色和纯白混合得浅色；和纯黑混合得深色；加入不同量的黑白色会得到不同的深浅效果。如果按照颜色的深浅将它们排列在一起，会呈现出渐变的节奏感（图5-1）。

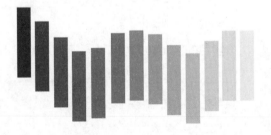

Figure 5-1　Color gradient
图5-1　色彩渐变

Chroma 纯度
This is the purity of a color in relation to gray, and can also be called saturation. If

the color is added to different amounts of gray, it reveals different levels of brightness or dullness. The greater the amount of gray, the duller the color, conversely the brighter; e.g. vivid blue and dusty blue, or grass green and olive green; although they can have the same hue and value.

纯度指和灰色相关的纯粹程度，也称为饱和度。颜色中加入不同量的灰色，会呈现出不同的鲜灰程度。加入的灰色越多，颜色越暗；反之颜色越鲜艳。例如，湖蓝和灰蓝、草绿和橄榄绿，但它们可以有相同的色相和明度。

Further reading: Relative terms

primary colors | red, yellow, and blue. These colors cannot be created by combining any other colors. Primaries are used to create all other colors.

secondary colors | green, orange and purple. These are created directly from a combination of two primary colors.

tertiary colors | yellow-orange, red-orange, red-purple, blue-purple, blue-green, and yellow-green. These are formed by mixing a primary and secondary colors.

tone | the level of tint or shade of the color.

patina | the surface or texture of the color; often associated with the aging process.

延伸阅读：相关术语

三原色　红、黄、蓝。不能用其他任何颜色调和出来。三原色用于创建其他所有颜色。

次生色　绿、橙、紫。用两种三原色直接调和出来。

复合色　黄橙、红橙、红紫、蓝紫、蓝绿、黄绿。由三原色和次生色调和出来。

色调　颜色明亮或灰暗的程度。

色泽　颜色的外表或质感，通常与颜色的老化程度有关。

5.1.2 Colors culture 色彩文化

Originally color itself does not have any special meanings, but with the social and cultural development some colors were given specific meanings. These meanings around color range from the natural, the historical, the religious, the political, the psychological, to the purely emotional. For example, when people see green, they may associate spring, trees and lawns; when they see red, they may associate flowers, flame and warmth; and blue, they may associate the ocean, coolness and calmness. The combination of orange and yellow may make people think about desserts or fruits, so in the United States this combination is often used in restaurants to psychologically produce feelings of hunger in customers.

色彩本身是没有什么特殊意义的，但是随着社会和文化的发展有些颜色被赋予了特定的含义。这些颜色的含义从自然、历史、宗教、政治、心理，到纯粹情感的都有。比如，人们看到绿色会联想到春天、树木和草地；看到红色会联想到花朵、火焰和温暖；看到蓝色联想到海洋、清爽和平静；橙和黄色的组合会使人想起甜品或水果，因此这种色彩组合在美国的餐馆中经常使用，使顾客在心理上产生一种饥饿感。

Color has some profoundly complex relationships with social ideas and culture, and the meanings of colors vary from culture to culture with uniquely defined symbolism. Red in Chinese culture conveys good luck and auspiciousness; whereas Western societies sense danger when using the color on road signs. Yellow in ancient China represents supreme power, but in Christian culture it signifies timidity and jealousy. White in Western countries is the symbol of virginity and often used for a bride's wedding dress, while in China it is often used on mourning occasions. Other colors that denote mourning range from yellow in Egypt, to black in America, blue in Iran, red in South Africa, and purple in Thailand.

色彩与社会观念和文化之间有着深刻复杂的关系，色彩含义因文化的不同而有所差异，并具有独特的象征意义。红色在中国文化中传递着幸运和吉祥；而在西方社会，红色用到路标上让人感觉到的是危险。黄色在中国古代代表着至高无上的权利，而在基督教文化中却带有胆怯和嫉妒的意味。白色在西方国家是纯洁的象征，经常用于新娘礼服中，而在中国它却常用于丧礼。其他表示哀悼的颜色从埃及的黄色到美国的黑色、伊朗的蓝色、南非的红色、泰国的紫色都有。

Some colors' meanings stem from history and still exist today. Blue was once worn by ancient Roman public servants, and has thus been appropriated to the police uniform today. In history purple was considered a royal color in many countries because of sumptuary laws enacted in ancient times that allowed only nobility to wear purple. Moreover, its dye was costly and was laboriously extracted from a seashell found in the Mediterranean.

有些颜色的含义源于历史，直到今天仍然存在。蓝色曾经被古罗马的公务员穿着，于是用在了今天的警察制服里。在历史上，紫色被很多国家看作是皇家的颜色，因为古代有禁奢令，只有贵族才能穿着紫色，它的染料非常昂贵，是从地中海的一种贝壳里千辛万苦提取出来的。

Further reading: colors and their associated meanings

white	purity, surrender, truth, cleanliness, innocence, simplicity, coldness
black	intelligence, mourning, power, formality, elegance, evil, death, slimness, classicism
gray	conservatism, respect, wisdom, old age, pollution, boredom, decay
red	passion, energy, love, danger, anger, revolution, wealth
orange	happiness, playfulness, desire, optimism, sweetness
yellow	joy, cowardice, greed, femininity, friendship, warning
green	nature, fertility, youth, growth, health, new beginning, rawness, security
blue	calmness, water, oceans, peace, coolness, depression, rationalism
purple	nobility, envy, mystery, confusion, pride, instability, gaudiness
brown	rusticity, tradition, heaviness, dullness, poverty, roughness, earth

延伸阅读：色彩及其相关含义

白色	纯洁、投降、真理、清洁、清白、朴素、寒冷
黑色	智能、哀悼、权力、礼节、优雅、罪恶、死亡、苗条、古典主义
灰色	保守主义、尊重、智慧、老年、污染、厌倦、衰退
红色	热情、活力、爱情、危险、愤怒、革命、财富
橙色	幸福、嬉闹、欲望、乐观主义、甜蜜
黄色	欢乐、怯懦、贪婪、柔弱、友谊、警告
绿色	自然、肥沃、青春、成长、健康、新的开始、生、安全
蓝色	冷静、水、海洋、平静、凉爽、沮丧、理性主义
紫色	高贵、嫉妒、神秘、困惑、骄傲、不稳定性、艳丽
棕色	乡土气息、传统、沉重、迟钝、贫穷、粗糙、土地

Social customs can also define the meanings of colors. The combination of red and green has long symbolized the Christmas holiday, so there are many shops using red and green decorations during Christmas time. Red, white and blue in many cultures represent patriotism, thus they often appear on national flags. Navy and white are usually used in marine themes, no matter in fashion or other fields' design.

社会习俗也能定义颜色的含义。红绿搭配一直以来是圣诞节的象征，所以在圣诞节期间很多商店用红绿色做装饰。红、白、蓝在许多国家代表着爱国主义，因此它们经常出现在国旗上。海军蓝和白色通常用在航海主题中，不管是服装还是其他类型的设计。

5.1.3　Color harmony theory 色彩调和原理

Color harmony represents a good visual sense of balance or unity, and combinations of colors that exist in harmony are pleasing to the eyes. Below are some variations of color harmonies based on a color wheel; knowing some color harmony theories can help a designer find the right color combination. But please keep in mind, these harmony methods are more a color relationship tool rather than a tool of selection. When creating a color scheme, a designer should also consider the design theme and inspiration; use color harmonies along with hue, value and chroma. Besides these, the designer also needs to understand how people might react to the palette on a psychological basis.

色彩调和呈现出很好的视觉平衡和整体感，搭配谐调的色彩组合会使人眼睛愉悦。下面是一些基于色环之上的色彩调和种类。了解一些色彩调和原理可以帮助设计师做出合适的色彩搭配，但是请记住，这些调和方法更多的是一种色彩关联工具，而不是选择工具。创建色彩方案时，设计师还应考虑设计主题和设计灵感；再结合色相、明度和纯度来运用色彩调和原理；除此之外，设计师还要了解人们对这种色彩搭配会产生什么样的心理反应。

Monochromatic color scheme 单色调和

The colors of this combination are in the same hue family but in different values (Figure 5-2), e.g. dark red, red and light red. The harmony of these colors is easily achieved, but the shades or tints should be differentiated, if not, it maybe a little monotonous.

这种组合里的色彩来自同一个色相家族，但明度不同（图5-2），比如深红、大红和浅红。这类颜色比较容易调和，但在深浅上要有所区别，否则效果可能会有些单调。

Figure 5-2　Monochromatic color scheme
图5-2　单色调和

Analogous color scheme 邻近色调和

Analogous color schemes use colors that are next to each other on the color wheel (Figure 5-3), e.g. red, violet and purple, or yellow, lime and green. They usually work together well and create a serene and comfortable visual result. These colors can not only be harmonized easily, but also be full of change.

色环上相邻颜色的调和（图5-3），如红色、紫罗兰色

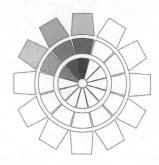

Figure 5-3　Analogous color scheme
图5-3　邻近色调和

和紫色，或黄色、黄绿色和绿色。它们通常可以很好地搭配，产生一种平静舒服的视觉效果。这些颜色不仅容易调和，而且富于变化。

Complementary color scheme 补色调和

Colors that are opposite to each other on the color wheel are complementary colors (Figure 5-4), e.g. red and green, yellow and purple, orange and blue. This color scheme has the strongest contrast and vibrant look. When creating it, designers would better not use all the colors at full saturation, or arrange them in similar proportions; otherwise they will be jarring.

Figure 5-4　Complementary color scheme

图5-4　补色调和

色环上相对的颜色是补色（图5-4），如红和绿，黄和紫，橙和蓝，这类调和具有最强烈的对比性和醒目的外观。在创建色彩方案时，不要让所有的颜色纯度都很高，或比例对等地排列，否则会不谐调。

Split-complementary color scheme 分离补色调和

It is a variation of the complementary color scheme; in addition to the base color, it uses the two colors adjacent to its complement (Figure 5-5). This color scheme has the same strong visual contrast as above, but has less tension. It should be uesd in some sportswear and swimsuit designs.

Figure 5-5　Split-comple-mentary color scheme

图5-5　分离补色调和

它是补色调和方案的变款，是把基础色和与它补色相邻的两种颜色进行调和（图5-5）。它具有同样强烈的视觉对比效果，但程度要小一点。可以用在运动装或泳装的设计当中。

Triadic color harmony 等差三色调和

In this scheme, the colors are evenly spaced around the color wheel (Figure 5-6). Triadic color scheme tends to be quite vibrant, even if some pale or unsaturated colors are used. This color scheme is very practical in fashion design.

Figure 5-6　Triadic color harmony

图5-6　等差三色调和

在色环上，这三色彼此间隔相同的距离（图5-6）。等差三色调和的效果往往比较明亮，即使是使用了一些灰色或低纯度的颜色。这种配色方案在服装设计中非常实用。

Double complementary color scheme 双重补色调和

The double complementary color scheme uses four colors arranged into two complementary pairs (Figure 5-7). This color scheme offers plenty of possibilities for color variation. The designer needs to pay attention to the balance between warm and cool colors when using it.

用两对互为补色的颜色进行调和（图5-7）。这种调和提供了色彩变化的多种可能性。使用时设计师需要注意暖色和冷色之间的平衡关系。

Figure 5-7　Double comple-mentary color scheme

图5-7　双重补色调和

5.1.4　Color match tips for fashion design 服装色彩搭配技巧

Firstly, the color blocks used in fashion design should not be arranged equally in size; that is to say, choose one color as the dominant color, a second color as the adjunctive color and the third as the ornamental color. The ornamental color occupies the smallest area but it is frequently used on some focus parts of a garment that can make these parts stand out. As figure 5-8, green is the dominant color; lime is the adjunctive color and red is the ornamental color.

首先，服装设计中使用的色块不应平均分配；也就是说，选择一种颜色作为主色，第二种颜色作为辅助色，第三种作为装饰色。装饰色面积最小，但经常用于服装的焦点位置以脱颖而出。如图5-8所示，绿色是主色，黄绿色是辅助色，而红色则是装饰色。

Secondly, to make sure there is some contrast in a color scheme. Designers can use different value or chroma of the colors to make them contrast. They can match the light, moderate and dark colors together, or the vivid, neutral and dull colors. For example, a khaki jacket and brown pants can be matched with a cream or citrus yellow sweater.

其次，确保色彩方案中有对比。设计师可以使用不同的明度和纯度使颜色具有对比性。可以把高明度、中明度和低明度色相搭配；或搭配高纯度、中纯度和低纯度色。比如，卡其色的上衣和棕色的裤子可以和奶油色或橘色的针织衫搭配在一起。

Figure 5-8　Color match for fashion design (YE's by YESIR Fall/Winter 2019 Fashion Show)

图5-8　服装色彩搭配（2019秋冬YE'S by YESIR 品牌发布）

Finally, to add some colored accessories as a complement or accent to brighten the outfit up. The color of a purse, hat, scarf, belt or some jewelry can all play a role of ornament.

最后，添加一些彩色的配饰作为补充或强调，使服装的整体色彩明亮起来。提包、帽子、围巾、腰带或首饰的颜色都可以起到点缀的作用。

5.2 Patterns for fashion design 服饰图案

5.2.1　Types of patterns for fashion design 服饰图案的种类

Patterns are also an important element for fashion design. They function as complements and make the fashion design more vivid. Patterns are created by color, lines and shapes; they come in an endless variety of forms (see the color page). The designs can be large or small, even or uneven, light or dark, spaced or clustered, muted or bold. Although patterns are only on some parts of a garment, they can also affect the whole design result.

图案也是服饰设计的一个重要元素。它可以起到锦上添花的作用，让服装设计更加生动。图案由颜色、线条和形状组成，形式非常丰富（见彩页）。图案设计可大可小，均匀或不均匀，明亮或黯淡，疏或密，柔和或醒目。虽然图案只是服装的局部，但也会影响到整体的设计效果。

In terms of styles, patterns for fashion design can be divided into concrete and abstract patterns. The former includes flora, animal, scenery, cartoon, and character patterns, etc. These patterns are usually used on women's and children's apparel and are rarely used on men's apparel. The latter mainly refers to the patterns that consist of points, lines and planes, like geometrics, stripes, checks, spots and random patterns, etc. They mostly appear on men's clothing, though sometimes they are also used to adorn women's clothes.

服饰图案根据风格可以分为具象图案和抽象图案。前者包括花纹、动物、风景、卡通和字符图案等。这类图案一般用在女装和童装身上，很少用于男装；后者主要指由点、线、面构成的图案，如几何、条纹、格子、散点和随机图案等。它们多用于男装，但有时也用来装饰女装。

Further reading: popular prints

spots	polka dots, random spots
stripes	pinstripes, chalk stripe
checks	tartan, plaids, gingham, houndstooth, Prince of Wales check
floras	small or large, dark flora, antique style or abstract art
geometrics	waves, zigzag, chevrons, herringbone, Art Deco style
novelty patterns	animals, unexpected items like aeroplanes or clowns
animal patterns	leopard, tiger, zebra, giraffe or cow
abstract patterns	splotches and scribbles, modern art style

延伸阅读：流行的印花

散点	波尔卡圆点、随机散点
条纹	细条纹、（深色地）白条纹
方格	苏格兰方格、格子、细方格、千鸟纹、威尔士亲王方格
花纹	小碎花或大花、暗花、仿古风格或抽象艺术
几何	波浪纹、之字纹、V形纹、人字形纹、装饰艺术风格
新奇图案	动物、意想不到的元素如飞机或小丑
动物图案	豹纹、老虎、斑马、长颈鹿或奶牛
抽象图案	墨点和涂鸦、现代艺术风格

5.2.2 Forms of patterns for fashion design 服饰图案的形式

How to use patterns to adorn the clothing? There are three main forms: pointed form, linear form and planar form. Pointed form patterns refer to a motif independently appearing on the garment. They highlight a certain part of the garment, and are easy to become the visual focus. Linear form patterns usually refer to the bidirectional continual patterns, and are suitable to adorn edges, such as the placket, bottom hem, neckline, etc. Planar form patterns are equivalent to the quadrilateral continual patterns and can adorn the whole garment or some parts of it. In many cases, designers use the patterns that are originally printed on the fabrics as their design elements. However, notice that large-scale patterns will have a feeling of expansion, so they are not suitable for plump people to wear.

怎样用图案来装饰服装？主要有三种形式：点状构成、线状构成和面状构成。点状构成图案指单独地出现在服装上的纹样。它强调服装的某一部分，形成视觉焦点。线状构成图案通常指二方连续图案，适合装饰边缘，如前襟、底摆、领口等。面状构成图案通常指四方连续图案，可以用于装饰服装整体或局部。很多情况下，设计师直接运用面料本来的印染图案作为设计元素。但需要注意，大花图案有膨胀感，不适合胖人穿着。

When a pattern is designed, how to transfer it to the clothing? The most common way is printing. There are two main technologies: screen printing and digital printing. Screen printing is a traditional technology and the cost is low, but sometimes it is restricted by the pattern colors' quantities. Now digital printing has become very common, which can "copy" the design motifs from computers to fabrics. As figure 5-9 shows, the pattern which is designed from a certain shell found on a beach is printed on the fabric. The disadvantage of digital printing is that the cost may be high if mass produced.

图案设计出来以后，如何用在服装上？最常见的方式就是印花。印花主要有两种技术：丝网印花和数码印花。丝网印花是一种传统的技术，成本低廉，但有时会受颜色套数的限制。现在数码印花已经非常普遍，可以把设计纹样从电脑上"复制"到面料上。图5-9中的图案是根据在海滩上发现的某种贝壳设计出来的，然后把它印在面料上。数码印花的缺点是如果大批量生产的话，成本会提高。

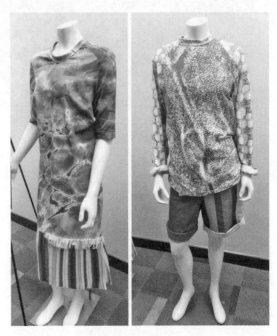

Figure 5-9   Digital printing (Designer: Emma Ptak)
图5-9   数码印花（设计：Emma Ptak）

Apart from printing, fashion design patterns can also be made by embroidering, painting, cutwork, sewing, appliqué, collage and some other crafts. These diversified making methods make the patterns be more liable to produce artistic effects. As figure 5-10 shows, the pattern is made using embroidering, and as figure 5-11 shows, the pattern is made with cloth sewing. Since fashion design is often considered a form of soft sculpture, the patterns can also be made into three-dimensional forms. Designers should not only design the pattern, but also think about how to make it.

除了印花，服饰图案还可以用刺绣、绘画、镂空、缝制、镶花、拼贴等其他工艺制作，这些多种多样的制作方法使图案更易于产生艺术效果。如图5-10所示，服饰图案是用刺绣工艺做的；再如图5-11所示，服装图案是用布料缝制的。服装设计通常被看作是软雕塑，所以服饰图案也可以做成立体的形式。设计师不仅要设计图案，还要考虑如何制作它。

Figure 5-10　Embroidering patterns
图5-10　刺绣图案

Figure 5-11　Cloth sewing patterns
图5-11　布料缝制图案

本章小结

- 色彩的三要素是色相、明度和纯度，色彩有一定的使用习俗和文化。
- 依据色环，色彩有一些调和方法。
- 服饰图案分为具象图案和抽象图案。
- 服饰图案有三种构成形式：点状构成、线状构成和面状构成。
- 服饰图案可以印在面料上，也可以通过其他工艺制作。

Exercises 课后练习

1. Please discuss with your partner how to match the colors in fashion design.

请与你的同伴讨论如何搭配服装色彩。

2. Please search for some fashion design pictures and describe the patterns used on them.

请搜寻一些时装图片并描述它们使用的图案。

3. Please discuss with your partner how to use the patterns to adorn the garments.

请与你的同伴讨论如何运用图案装饰服装。

Colors 色彩

salmon pink 肉粉色

coral 珊瑚色

peach 桃粉色

orange 橙色

red 大红色

pink 粉红色

magenta 洋红色

rose red 玫瑰红

ruby red/crimson 深红色

burgundy 酒红色

baby blue/pastel blue 浅蓝色

cyan 蓝青色

vivid blue 湖蓝色

dusty blue 灰蓝色

sapphire blue 宝石蓝

dark blue 深蓝色

navy blue 海军蓝

purplish blue 藏青色

liac 丁香色

lavender 淡紫色

mauve 浅紫红色

purple 紫色

violet 紫罗兰色

plum 紫红色

modena 深紫色

lime 黄绿色

pea green 豆绿色

kelly green 鲜绿色

aqua 水绿色（浅绿色）

olive green 橄榄绿

grass green 草色绿

emerald 宝石绿

deep green 深绿色

dark green 墨绿色

white 白色

off-white 本白色

cream 奶油色

ivory 象牙白

taupe 灰褐色

gray 灰色

charcoal 炭灰色

black 黑色

lemon yellow 柠檬黄

vibrant yellow 亮黄色（迎春黄）

beige 米色

citrus yellow 橘黄色

honey yellow 蜜黄色

khaki 卡其色

ochre 赭色

brown 褐色

Patterns 图案

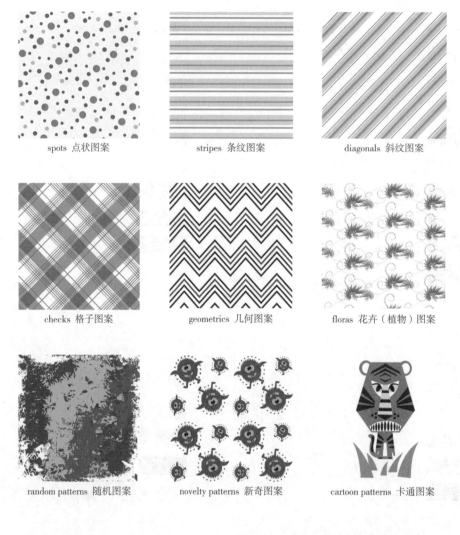

spots 点状图案

stripes 条纹图案

diagonals 斜纹图案

checks 格子图案

geometrics 几何图案

floras 花卉（植物）图案

random patterns 随机图案

novelty patterns 新奇图案

cartoon patterns 卡通图案

quadrilateral continual patterns
四方连续图案

bidirectional continual patterns 二方连续图案

Fabric for Fashion Design
服装面料

课题名称： Fabric for Fashion Design

课题内容： 服装面料

课题时间： 2课时

教学目的： 让学生学习并掌握服装面料的种类，主要面料的性能、特点，如何选择面料，以及面料表面装饰方法的叙述和表达。

教学方式： 结合PPT与音频等多媒体教学课件，以教师课堂讲解为主，学生随堂练习与小组讨论为辅。

教学要求： 1. 掌握相关词汇。

2. 能够用英语描述出主要服装面料的种类及特点，能列举出面料表面装饰方法的名称。

课前（后）准备： 课前预习关于服装面料和面料装饰方法的基础知识，浏览课文；课后记忆单词，并复述文中内容。

6.1 Natural and chemical fibers
天然纤维和化学纤维

6.1.1　Natural fiber–plant fibers 天然纤维：植物纤维

Natural fibers are derived from organic sources; these can be divided into plant sources which are composed of cellulose, and animal sources which are composed of protein.

天然纤维来源于有机物质，可分为由纤维素构成的植物纤维和由蛋白质构成的动物纤维。

Cotton 棉

Cotton is the most commonly used natural fiber. Cotton possesses the qualities of comfort and natural absorbency, which is useful in hot climates as it can absorb moisture and dry off easily. It can embody, depending on processing, either cool or warm aspects that make it truly trans-seasonal. For instance, cotton can be made into cotton coats in winter for its softness and high warmth retention property; and made into T-shirts in summer because of its good breathability. Cotton is liable to shrink, so it is better to be pretreated with shrink-resistant substances before it is made into fabrics.

棉是最常用的天然纤维。棉穿着舒适，具有天然的吸水能力，在气候炎热的条件下很实用，因为它吸湿易干。根据不同的加工方法，它可以冷暖两用，真正具有跨季节性。例如，因轻盈保暖，冬天可以做成棉袄；因透气性好，夏天可以做成T恤。棉布很容易缩水，所以在做成面料之前最好先用防缩水物质做预处理。

Linen 亚麻

Linen has been highly valued for many centuries for its refined and subtle feel. Its cool, absorbent properties are well recognized the world over and are unparalleled by any other natural fiber. Its natural creasing lends clothing a unique charm; its inherent anti-static properties make it fall away from the body, and undulate in response to the wearer's movement. Linen is also liable to shrink as cotton, but blending with some synthetic fibers will improve this.

亚麻因其精致而细腻的手感几百年来一直都很受推崇。它凉爽、吸湿的特性广为人知，其他天然纤维都望尘莫及。它自然的褶皱给服装增添了独特魅力，它固有的防静电功能使服装不会紧贴身体，且随人体的活动而自然起伏。亚麻和棉一样也容易起皱，但和一些合成纤维混纺以后会得到改善。

Bamboo 竹

Nature bamboo fiber has both soft and smooth properties; therefore it has earned the description "cashmere from plants". Its fiber structure allows it to wick away and evaporate perspiration instantly, keeping the wearer drier and cooler. Bamboo fiber is known as an ideal choice for active sportswear fabrics. It is said that the fiber will keep the wearer one or two degrees cooler in summer than other natural fibers, which is perhaps why it is marketed in some Asian countries as "air-conditioning" fiber.

天然竹纤维具有柔软光滑的特点，因此赢得了"植物羊绒"的美誉。它的纤维结构使它可以快速吸收和蒸发人体的汗液，让穿着者感觉干爽和清凉。竹纤维被认为是制作竞技运动服装面料的理想选择。据说在夏天穿着时，比起其他天然纤维可以使穿着者的体表温度低1—2度，这也许就是一些亚洲国家将"空调"纤维作为它的营销策略的原因。

6.1.2　Natural fiber–animal fibers 天然纤维：动物纤维

Wool 羊毛

The natural properties of wool make it warm, resilient, flexible, absorbent and moldable. Wool fiber's unique structure lends it a good insulating quality which can maintain more warmth; thus, it is usually used to make scarves, sweaters, gloves, etc. It can also absorb up to 30 percent of its bulk weight in moisture vapor without feeling wet; this makes it still feel warm while wet. Wool can embody many diverse aspects, which can be satisfyingly soft or rugged.

羊毛的天然特性是保暖、有回弹力、柔韧、吸湿、可塑性强。羊毛纤维的特殊结构使它具有很好的绝缘性能，可以保持更多的热量，所以经常用来制作围巾、毛衫和手套等；它还可以吸收高达它体积容量30%的湿气而不让人感到潮湿，这使它在潮湿的状态下仍然让人感觉到温暖。羊毛的可塑性强，可柔软，亦可硬挺。

Cashmere 羊绒

Cashmere is produced from a fine downy undercoat of goats. Its name is synonymous with luxury and is described as "soft gold". Cashmere from Mongolia and the Inner Mongolian Autonomous Region of China is of the best quality. Much of this is due to the extreme climate, which encourages the goats to grow a finer, denser under-hair. Cashmere fibers are very soft, sleek, comfortable and much warmer than wool.

羊绒是从山羊身上细密柔软的底层绒毛中产生出来的。它是奢侈的代名词，被称为"软黄金"。产自蒙古国和我国内蒙古自治区的羊绒品质最好。这主要是因为适应那里极端的气候，促使山羊长出更精细浓密的底绒来。羊绒纤维非常柔软、光滑、舒适，比羊毛更加保暖。

Silk 丝

Silk is a protein fiber harvested from the cocoon of the silkworm. The cocoon is made from a continuous thread that is produced by the silkworm which it wraps around itself for protection. Silk possesses a magnificent, shimmering richness that can be woven into sumptuous satins, brocades, and jacquards. It is so luxurious and precious that its price has, at times, exceeded that of gold. Silk is widely perceived to be the most beautiful and elegant of all the natural fibers, and even after more than a century of attempting to provide a man-made substitute, no single synthetic fiber has come close to replicating either the luster or the supple, liquid drape of silk.

丝是一种蛋白质纤维，是从蚕茧中提取出来的。蚕茧是由桑蚕吐出的一根长丝形成，蚕为了保护自己用蚕茧将自己包裹起来。丝纤维光华夺目，可以织成华美的绸缎、织锦和提花织物。它如此奢华珍贵，以至于有时重金难求。丝在所有天然纤维中被普遍认为美丽高雅，即使经过了一个多世纪的尝试，人们想寻找一种人工代替品，但不论是光泽，还是柔软、顺滑的垂感，还没有一种合成纤维能够完全取代它。

Further reading: natural fiber fabrics

cotton	corduroy, cotton sateen, organdy, canvas, seersucker
linen	linen, batiste
wool	gabardine, wool felt, melton, flannel
silk	georgette, organza, chiffon, voile, crepe, velvet

延伸阅读：天然纤维面料

棉	灯芯绒、棉缎、蝉翼纱、帆布、泡泡纱
亚麻	亚麻、细亚麻
毛	华达呢、羊毛毡、麦尔登、法兰绒
丝	乔其纱、欧根纱、雪纺、巴厘纱、真丝绸、天鹅绒

6.1.3 Chemical fibers 化学纤维

Chemical fibers can be divided into two principal categories: regenerated and synthetic. Regenerated fibers are regenerated from natural sources and can be subdivided into plant cellulose and bio-engineered fibers. Plant cellulose fibers are regenerated from such plant sources as wood; bio-engineered fibers describe a new generation of fibers that bridge the gap between fiber science and polymer science, and may use proteins, sugars, or starches as their starting point. Synthetic fibers cover fibers made from chemicals and derived from such fossil fuels as oil and coal. Moreover, inorganic fibers are also categorized as chemical fibers.

化学纤维主要分为再生纤维和合成纤维两大类。再生纤维是从天然来源中再生而成，可以再细分为植物纤维素纤维和"生物工程纤维"。植物纤维素纤维是从像木材一样的植物资源中再次生成的；生物工程纤维指的是一种衔接了纤维和高分子科学的新生代纤维，它可能取材于蛋白质、食糖或淀粉。合成纤维是用从石油和煤炭等矿石燃料中提取的化学物质制成的纤维。另外，无机纤维也属于化学纤维。

Regenerated fibers 再生纤维

Viscose was one of the first regenerated fabrics to be developed and is arguably the most popular cellulose fabric to date. It shares some properties with silk or cotton, depending on different processing techniques. Viscose is absorbent and has a soft and smooth touch, but is not especially durable. Different chemicals and processes are used in the production of viscose, so each has a different name, such as lyocell, modal and tencel, which are new environmentally friendly regenerated fabrics.

黏胶是最早研发的再生纤维面料之一，可以说是迄今为止最受欢迎的纤维素纤维面料。根据不同的生产工艺，它可以具有丝或棉的特性。黏胶吸湿性好，手感柔软光滑，但不是很耐用。用不同的化学成分和工序生产的黏胶有不同的名字，如莱赛尔、莫代尔和天丝，这些都是新型的环保再生纤维面料。

Corn fiber, soybean fiber and milk fiber are all classified as bio-engineered fibers. The corn fiber has low flammability, and is more resistant to ultraviolet (UV) light, and is very hydrophilic. The soybean fiber fabric drapes well and is often considered to be as smooth as cashmere, while its luster can resemble that of silk. Milk fiber fabrics have certain wool-like characteristics, and so the preferred fibers for blending are wool and cashmere.

玉米纤维、大豆纤维和牛奶纤维都属于生物工程纤维。玉米纤维不易燃，抗紫外线能力强，且亲水性好。大豆纤维面料的垂感好，被公认为具有羊绒般的光滑手感和丝绸般的光泽。牛奶纤维面料具有一定的羊毛特性，所以与之混纺的纤维首选羊毛和羊绒。

Further reading: the term of viscose

In continental Europe it is known as viscose, in Great Britain it may be referred to as viscose rayon, and in the United States the preferred popular term is rayon. By contrast, the European designation, "viscose", is more pragmatic, being descriptive of the processing method used to produce the fiber: production of a highly viscous solution.

延伸阅读：“黏胶”的术语

在欧洲它被称为"viscose"，在英国会被称为"viscose rayon"，而在美国较流行的叫法是"rayon"。相比之下，欧洲的叫法更为实用，因为它描述了黏胶的加工工艺，那就是用高黏度的溶液进行生产。

Synthetic fibers 合成纤维

Polyester is the most widely used synthetic fiber. It is an easy care and crease-resistant fiber, and to some extent, low-cost. It can be made from clean plastic drink bottles. Polyester does not absorb moisture well, but if blended with natural fibers its ability to absorb moisture increases.

涤纶是应用最广泛的合成纤维。它好打理，抗起皱，从某种程度上讲，是一种价格低廉的纤维，它可以从干净的塑料饮用瓶中提取。涤纶的吸湿性差，但若与天然纤维混纺将会改善这一缺点。

DuPont of the US is the most famous and one of the earliest companies inventing and producing nylon. Nylon (polyamide) is a strong, smooth, lightweight fiber to which dirt cannot cling easily to its surface. Nowadays nylon is also synonymous with stockings.

美国的杜邦公司是最著名且最早发明和生产尼龙的公司之一。尼龙（锦纶）是一种结实、光滑、质轻的纤维，且不易沾尘。如今尼龙还是丝袜的代名词。

Acrylic has the look and handle of wool which is usually used to make heavy sweaters, artificial furs and fluffy toys.

腈纶具有羊毛般的外观和手感，经常用来制作厚毛衫、人造皮草和毛绒玩具。

Spandex fibers are elastic fibers that can be used in a variety of garments to improve fit and function. Spandex is differentiated from other fibers by its high stretch and recovery. The best known spandex fiber brand is lycra produced by DuPont.

氨纶是一种弹性纤维，可用于各种服装中以改善它们的合体性和功能性。氨纶不同于其他纤维，它具有很好的伸展和回弹能力。杜邦公司生产的"莱卡"是最著名的氨纶商标。

6.1.4　Blended fabrics 混纺面料

Fabrics today are often formed of a blend of natural and chemical fibers. A fabric can therefore have the best properties of each fiber. For instance, with a polyester/cotton mixed fabric, the cotton lends breathability and the polyester is crease-resistant; with a cotton/lycra blend, the lycra will lend elasticity to the cotton fabric.

当今的面料通常是由天然纤维和化学纤维混纺而成的，这可以使不同类型的纤维在性能上取长补短。例如，涤棉混纺面料，棉透气性好，涤纶不易起皱；棉和莱卡混纺，可以增强棉的弹性。

Further reading: man-made fibers category

regenerated fibers	rayon/viscose, tencel, lyocell, corn fiber, soybean fiber, milk fiber, chitin fiber, alginate fiber
synthetic fibers	polyester, polyamide (nylon), acrylic, spandex
Inorganic fibers	glass fiber, metallic fiber, ceramic fiber

延伸阅读：化学纤维的种类

再生纤维	黏胶、天丝、莱赛尔纤维、玉米纤维、大豆纤维、牛奶纤维、甲壳素纤维、海藻纤维
合成纤维	涤纶、锦纶（尼龙）、腈纶、氨纶
无机纤维	玻璃纤维、金属纤维、陶瓷纤维

6.1.5　How to choose the fabrics 如何选择面料

Firstly, choosing a fabric should consider the way it looks or its aesthetic result, in other words, the visual feel that they bring to people, including color, pattern and texture, etc.

首先，选择面料要考虑它的外观或美学效果，也就是它带给人们的视觉感受，包括颜色、图案和肌理等。

Secondly, choosing a fabric should consider its weight and handle. The weight and handle of a fabric will affect the silhouette of a garment, giving it shape and form and allowing it to drape. The handle of a fabric is referring to how it feels in hands, its tactility, and also the way it drapes or hangs from the hand.

其次，选择面料要考虑其重量和手感。面料的重量和手感会影响服装的廓型，赋予服装造型和形态，并使之下垂。面料的手感是指它拿在手里的感觉，它的触感，以及从手中下坠悬垂的方式。

Thirdly, choosing a fabric should also consider its performance. For example, jeans must be comfortable, durable and wear-resistant, and therefore denim is the perfect fabric for them. A T-shirt would be best made of a breathable fabric; possibly 100% cotton-knitted jersey is the ideal choice.

再次，选择面料还要考虑它的内在性能。比如，牛仔裤必须要舒适、耐磨、耐穿，斜纹粗棉布是制作牛仔裤的最佳面料。T恤衫最好用透气性好的面料制作，比如100%的纯棉针织汗布是其理想的选择之一。

Lastly and probably of most importance modern-day designers should consider the ethical impact that fabric has on the environment. When choosing a fabric, designers should question whether the fiber is organic in origin, whether the dyes used to color the yarn are eco-friendly or whether the factory that the fabric is made in meets trade standards. Sometimes they should also consider if it is necessary to use real fur or leather in designs to make the artwork perfect. Besides above, there are many other ethical questions that a designer needs to consider.

最后且可能最重要的是，现代设计师应该考虑面料对环境带来的伦理影响力。选择面料时，设计师应该询问这种纤维的来源是否有机，染制纱线的染料是否环保，生产面料的厂家是否符合行业标准。有时他们还应考虑，在设计时是否有必要使用动物皮草或皮革才能使艺术作品达到完美。除此以外，还有很多其他的伦理问题需要设计师们考虑。

6.2 Peculiar materials for fashion design
特殊服装材料

In addition to the common fabrics, there are still some peculiar materials that can be used in fashion design. Although these materials cannot be directly worn in daily life, they usually have strong artistic effects and visual impact, which express well the designers' ideas and individuality. There are many costumes made of peculiar materials frequently appearing on the runway. For example, the Spanish designer Paco Rabanne used metal sheets in the miniskirt design; the British designer Gareth Pugh used metal to make the tassels of dresses; and Japanese designer Yohji Yamamoto used wood panels to form a dress. In a chocolate salon in Paris the designer used chocolate to make the costumes and accessories. Besides these, there are some other designers who have used leaves, feathers, plastic, bamboo, ropes, etc. to design their artworks. These not only provide us with a unique visual sense, but also challenge the traditional fabrics for fashion design (Figure 6-1).

Figure 6-1　A dress made of paper

图6-1　纸做的服装

　　除了普通面料，还有一些特殊材料可以用于服装设计。虽然这些材料不能直接在生活中穿着，但它们往往具有强烈的艺术效果和视觉冲击力，很好地表达了设计师的想法和个性。时装舞台上经常会出现用特殊材料制作的服装。例如，西班牙设计师帕高·拉巴纳曾用金属片设计迷你裙；英国设计师加雷斯·皮尤用金属制作礼服的流苏；日本设计师山本耀司用木板制作连衣裙。在巴黎的一次巧克力沙龙上，设计师用巧克力制作服装和饰品。除此以外，还有的设计师用树叶、羽毛、塑料、竹子、绳索等材料塑造他们的艺术作品。这不仅给我们带来了独特的视觉感受，也向传统的服装面料提出了挑战（图6-1）。

6.3 Fabric surface decorations
面料表面装饰

6.3.1　Smocking 抽褶

Smocking is to gather a fabric into some dense and regular puckers. It will work better on a solid fabric (Figure 6-2). Designers need to first mark the grids or dots in place and then gather the cloth through a single thread.

　　抽褶法是把面料抽成紧密而又有规律的皱褶（图6-2）。最好用素色面料制作。设计师需要先在恰当的位置画好方格或圆点，再用一根线把布料抽起来。

6.3.2　Cutwork 镂空

Cutwork is to cut away some areas of a fabric according to a certain pattern, and the raw edges are locked by stitches to prevent from fraying (Figure 6-3). Cutwork can also be made with a laser. Laser-cut not only works quickly, but also guarantees the designed patterns to be more precisely "copied" onto the fabric. During this process the raw edges are sealed with heat simultaneously.

镂空就是按照一定的图案剪掉面料上的一部分，再缝锁毛边，防止脱散（图6-3）。镂空也可以用激光制作，激光雕刻不仅迅速，还能保证设计的图案更加准确地"复制"到面料上，同时在此过程中将毛边用高温封住。

Figure 6-2   Smocking (Maker: Zheng Danli)
图6-2   抽褶（制作：郑丹莉）

Figure 6-3   Cutwork (Maker: Li Wenjun)
图6-3   镂空（制作：李文军）

6.3.3   Folding 折叠

A fabric is folded into kinds of shapes just as paper folding (Figure 6-4), and is fastened on a background cloth; this craft can create some three-dimensional effects.

将面料像折纸那样折成各种形状，再固定于底布上（图6-4），这种工艺可以做出一些立体效果。

6.3.4   Burning 燃烧

Most fabrics' fire points are low. If some areas of a fabric are ignited with high temperature, they will be charred. When these areas contrast with the remaining parts of the fabric, it will produce another effect (Figure 6-5).

多数面料的燃点都很低。若将面料上的有些部分用高温点燃的话，会被烧焦。烧焦的部分和剩下的部分形成对比，会产生另外一种效果（图6-5）。

Figure 6-4　Folding (Maker: Lu Yingying)
图6-4　折叠（制作：陆影影）

Figure 6-5　Burning (Maker: Yang Dengshun)
图6-5　燃烧（制作：杨登顺）

6.3.5　Discharge printing 拔染

A bleaching agent is painted onto a fabric to remove some of the color. The removed areas accompanying the remaining colors may create a novelty effect (Figure 6-6).

将漂白剂涂在布料上以去除一些颜色。去除颜色的部分和剩余的颜色结合在一起，可能会产生新奇的效果（图6-6）。

Figure 6-6　Discharge printing (Maker: Wang Lixiao)
图6-6　拔染（制作：王立晓）

6.3.6　Resist dyeing 防染

It refers to various techniques of patterning on a fabric by preventing dye reaching some certain areas of it (Figure 6-7). Batik and tie-dye are the two common techniques. Batik is to paint wax onto a fabric to form a motif, and then dip the fabric into dye liquor. The waxed areas will not take the dye, leaving uncolored motifs against a colored background; meanwhile they can produce natural ice-crack patterns. Tie-dye is to tightly bundle or tie some certain parts of the cloth to prevent dye liquor from permeating.

防染指防止染料染于面料上的某些区域，而形成图案的各种工艺（图6-7）。蜡染和扎染是两种常用的方法。蜡染是用蜡在布料上画上纹样，再浸入染液。着蜡的地方不会被染色，从而在染色的底布上形成白色的花纹，同时产生自然的冰裂纹。扎染是将布料上的特定区域紧紧捆住或打结，以阻止染液的渗透。

Figure 6-7   Resist dyeing (Maker: Students of Qingdao University)
图6-7　防染（制作：青岛大学学生）

6.3.7   Appliqué 嵌花

Appliqué covers a wide number of methods that add pieces of fabrics to a background cloth, usually for decorative purposes. The shapes can be flowers, animals or geometries; two-dimensional or three-dimensional (Figure 6-8). The pieces can be fastened by stitching, or pasting on with special glue.

嵌花的方法有很多，是在底布上添加一些面料裁片，通常是为了装饰。其造型可以是花朵、动物或几何形，平面或者立体（图6-8）。这些裁片可以缝合固定，或者用特制的胶水粘住。

6.3.8   Coiling 盘绕

Figure 6-8   Appliqué (Maker: Guan Wanting)
图6-8　嵌花（制作：关婉婷）

It refers to coiling ropes, threads or cloth strips into some artistic patterns, and then fastening onto a background cloth (Figure 6-9). The fastening methods are similar to appliqué.

将绳子、线或布条盘成艺术图案，固定于底布之上（图6-9）。固定的方法与嵌花类似。

6.3.9　Crocheting 钩编

Crocheting is the method of creating fabrics from threads or ropes. Designers use a crochet needle to interlock the loops together; during this process the designed patterns are also made (Figure 6-10).

钩编是一种用线或绳创造织物的方法。设计师用钩针将线圈套锁在一起，花样也在此过程中钩出（图6-10）。

6.3.10　Embroidering 刺绣

Embroidering is to sew some decorative patterns on a background cloth with threads of various colors (Figure 6-11). The materials,

Figure 6-9　Coiling (Maker: Zhang Qingyu)
图6-9　盘绕（制作：张晴雨）

techniques and stitches may differ according to the design. The traditional embroidering craft is time-consuming and needs designers' patience, but nowadays some of the work can be done by a machine. Embroidering can lend fabrics an exquisite effect.

刺绣是在底布上用各种颜色的线缝出装饰图案（图6-11）。其材质、工艺和针法可能根据设计而有所不同。传统的刺绣工艺非常耗时，需要设计师有耐心，但是现在有些工作可以用机器完成。刺绣可以为面料增添精细的效果。

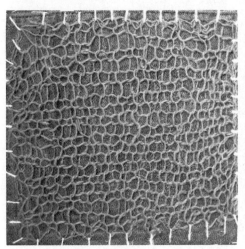

Figure 6-10　Crocheting (Maker: Xu Pei)
图6-10　钩编（制作：徐沛）

Figure 6-11　Embroiderin (Maker: Fan Chunxiao)
图6-11　刺绣（制作：范春晓）

6.3.11　Beading 珠绣

Beading is in essence embroidering with beads; each bead is fastened to the cloth by sewing (Figure 6-12). The beads can be glass beads, pearls, sequins, or made of crystals, plastic, wood, bones, and so on. Beading adds texture to the fabric, lending a delicate and shiny effect to it.

珠绣实质上就是用珠子进行刺绣，每一颗珠子都是用针缝的方法固定在布料上（图6-12）。珠子可以是玻璃珠、珍珠、亮片，或用水晶、塑料、木头、骨头等材料做成。珠绣给织物增添了质感，并赋予其精美、闪亮的效果。

Figure 6-12　Beading (Maker: Li Yutong)
图6-12　珠绣（制作：李禹潼）

6.3.12　Hand Painting 手绘

This is to paint on cloth with special pigments and tools. The methods are similar to painting on paper, but the pigments should not be painted very thick.

这是用特殊的颜料和工具在布料上作画。画法和在纸上画相似，但颜料不能画得太厚。

6.3.13　Quilting 绗缝

Quilting is to sew two or more layers of cloth together with a layer of batting sandwiched between them, in order to create a three-dimensional effect. Sometimes the stitches are some decorative elements as well.

绗缝是将两层或多层的布料缝合在一起的工艺，其中夹了一层棉絮，以创造出立体的效果。有时缝线也是一种装饰元素。

6.3.14　Heat-transfer printing 热转移印花

A piece of cloth and special pre-printed paper are pressed between heated panels, transferring the dye from paper to the cloth by heat (Figure 6-13). This is a low-cost producing process which is suited for short runs.

将布料和一种特制的预先印好图案的纸放在加热板之间，通过加热将染料从纸上转印到布料上（图6-13）。这是一种低成本的生产工艺，适合小批量印染。

Figure 6-13　Heat-transfer printing
（ Maker: Jiao Youzhu ）
图6-13　热转移印花（制作：焦幼竹）

本章小结

- 纤维分为天然纤维和化学纤维：天然纤维主要包括植物纤维和动物纤维；化学纤维主要包括再生纤维和合成纤维。
- 服装除了用普通面料制作外，还可以用特殊材料制作。
- 对服装面料进行表面装饰，可以产生艺术效果。

Exercises 课后练习

1. Create your own "Fabric Library". 创建"面料库"。

Please collect some fabric pieces and paste them on the cardboard. And then write down their names and features correspondingly.

请收集一些面料小样粘在卡纸上，并一一对应写上它们的名称和特点。

2. Please use multiple methods of fabric surface decorations that illustrated above to design your fabric art works, and exchange with your classmates.

请运用上面介绍的多种面料表面装饰方法，来设计你的面料艺术作品，并和你的同学进行交流。

Chapter 7
第七章

Accessories
配 饰

课题名称： Accessories

课题内容： 服装配饰

课题时间： 2课时

教学目的： 让学生学习并掌握配饰的作用，各种配饰的类型、发展历史和设计搭配。

教学方式： 结合PPT与音频等多媒体教学课件，以教师课堂讲解为主，学生随堂练习与小组讨论为辅。

教学要求： 1. 掌握相关词汇。

2. 能够用英语描述出各种配饰的造型特征和搭配方法。

课前（后）准备： 课前预习各种配饰的种类、特征和发展历史；浏览课文；课后记忆单词，并复述文中内容。

7.1 The role of accessories and their position in fashion industry
配饰的作用及其在服装业中的地位

Accessories are an indispensable part of the fashion family; they include the practical articles and adornments that accompany the clothing. Each one has a specific purpose, such as protective, functional or ornamental. Accessories play a key role in the presentation of the outfit that they accompany, and can also say something to us of the person who wears it.

配饰是时尚家族中不可缺少的一部分，它包括搭配服装的实用物品和装饰品。每件配饰都有其特定的用途，如保护性、功能性或装饰性。配饰在服装搭配中起着关键的作用，还能传达出关于佩戴者的一些信息。

The main role of accessories is to complete a man, woman, or a child's outfit. For example, a lady in a dull-color uniform will be more attractive if she wears a vivid-color scarf, or carries a purse in contrasting colors. Accessories can complement the outfit's color if the harmony is monotonous. Accessories usually act as a finishing touch which can highlight a person's temperament, e.g. a lady wears a shiny brooch or earrings; or a gentleman wears a patterned tie. Additionally, accessories are also an important element of expressing people's individuality.

配饰的主要作用是使男女装或童装更加完善。例如，一位身着暗色调制服的女士，如果戴一条亮色的丝巾或配一个撞色的提包，会更有魅力。如果服装色彩较单调，配饰可以起到补充色彩的作用。配饰通常用作衬托气质的点睛之笔，如女士佩戴闪亮的胸针或耳环；男士佩戴有花色的领带。此外，配饰还是人们表达个性的重要元素。

Accessories occupy a place of ever-increasing importance in the modern fashion industry and, in many cases, are the most profitable. Accessories have become an important way for most consumers to have a bit of luxury. Some luxury brands have transformed accessories into their best means of reaching the general public, since their accessories offer the satisfaction of long-cherished aspirations of the public.

配饰在现代服装业中占据着越来越重要的位置，而且在很多情况下，是最能产生利润的部分。配饰已经变成大多数消费者能够拥有奢侈品的重要方式。一些奢侈品牌已经将配饰变成走入大众的最佳途径，因为他们的配饰满足了大众梦寐以求的愿望。

The majority of brands that are recognized mainly for manufacturing clothes have been extending their range of accessories, in answer to an increasing demand on the part of the consumer. Their objective, in addition to increasing the number of sales, it is to be able to reinforce their presence in the fashion market. Some brands have accessories as the basis for their turnover and have gained a strong position in the fashion market thanks to them. Accessories can act like pieces of merchandising or complementary objects to items of clothing. If the accessories are distinctive and their design matches the style of the company, they can become a key tool in constructing the brand image. This fact encourages the companies of these brands to complement the accessories in creating garment collections.

多数以生产服装为主的品牌都拓展了配饰业务，以满足消费者日益增长的需求。他们的目标除了提高销售额外，还能增强自身品牌在服装市场中的存在感。一些品牌将配饰作为取得营业额的基础，而且配饰也帮他们在市场中占有强有力的地位。配饰可以以单件商品或服装配件的形式出现。如果配饰很有特色且设计符合公司风格的话，便能成为塑造品牌形象的关键要素，这一点促使这些服装品牌在系列创作中都纷纷加入了配饰。

7.2 Categories of accessories
配饰的种类

7.2.1　Bags 包袋

The bag or the purse (BrE handbag), a common accessory used to keep objects for daily use, is both practical and decorative. It varies in shape and incorporates details according to its use as determined by designers (Figure 7-1). In the beginning it was a typically feminine accessory, but changing customs have turned it into a normal part of the male wardrobe as well.

包袋，是用来盛放日常用品的常用饰物，既有实用性又有装饰性。现代包袋造型各异，细节多样，皆由设计师根据其用途决定（图7-1）。它最初是典型的女性饰品，但随着生活习俗的变迁，也已成为男人衣橱中很常见的一部分。

The first purses in the history of fashion were deemed "ridiculous", small pieces that served to keep objects for personal care, but became fashionable by the end of the 18th century. New ideas governing dress early in the 20th century did not provide much space in which to keep things, so the purse became the perfect ally for ladies and gentlemen of the time. From then on, every decade has seen specific styles appear. The 1920s marked a clear tendency to use rectangular purses, whereas in the 1950s purses with long straps appeared, in response to the need to have the hands free. At the end of the 20th century , sports

shell-shaped purse 贝壳包

frame purse 铰扣包

clutch purse 手握包

sling bag 单肩包

barrel purse 圆筒包

backpack 双肩背包

hobo purse 肚囊包

flap purse 翻盖包

dorothy purse 束口提包

briefcase 公文包

tote purse 大型提包

Figure 7-1 Bags design variation
图7-1 包袋款式设计

designs have become popular, which are generously sized, perfect for the new rhythm of urban life.

最早的包袋出现时被认为是"可笑"的，用如此一种小物件来盛放私人用品，但是到了18世纪末包袋就变得非常流行了。20世纪初新的着装理念使服装没有足够的空间盛放物品，包袋便成为当时男士和女士们着装的完美伴侣。从那时起，每十年都有特定的款式出现。20世纪20年代使用方形包成为明显的趋势；而在20世纪50年代长肩带背包出现了，可以将双手解放出来。到20世纪末，运动风格的包袋流行，它们的尺寸较大，非常适合都市生活的新节奏。

There are a variety of purse designs in modern fashion: the shape may be large or small, flat or boxy; the color may be single or multicolored. The most common material used to make purses is leather, usually from the calf or goat, or even from the pig, python or the crocodile. However, fabric is a more and more frequent resource now, like cotton, silk, canvas, imitation leather and PVC.

现代的包袋设计可以说是千姿百态：形状或大或小，或扁或方；颜色或单色或多色。制作包袋最常用的材料是皮革，一般来自小牛皮或山羊皮，乃至猪皮、蟒皮或鳄鱼皮。不过现在越来越多的包袋用织物面料制作，像棉布、丝绸、帆布、仿皮革和PVC（聚乙烯）材料等。

7.2.2　Footwear 鞋类

The design of footwear, in terms of form, has hardly evolved; the majority is reduced to a certain number that relates to timeless patterns (Figure 7-2). However, footwear creation gives endless possibilities of combinations in terms of materials and colors. On some special occasions, such as fashion shows, designers may surprise us with models demonstrating a large dose of the theatrical.

鞋的设计，从造型上讲，不会有太大的突破；大多数现代鞋的设计都简化为某些恒久不变的经典款式（图7-2）。但在材料和色彩搭配方面，鞋子的创意又具有无限的可能性。在时装发布会等一些特定场合，设计师也许会通过模特展示的大量戏剧性鞋子设计带给我们惊喜。

Sandals were the first documented model of unisex footwear that appeared in the history of fashion (Figure 7-3) . In the middle of the 16th century, the first footwear that incorporated a small heel appeared. However, at that time heeled shoes were not mainly designed to be worn but to be admired. This footwear symbolized the social prestige and opulence of the person who wore them.

round toe pump 圆头高跟鞋   sandal 凉鞋   open back pump 露跟高跟鞋

stiletto 细高跟鞋   wedge 坡跟鞋   peep-toe pump 鱼嘴鞋

flip flop 人字拖鞋   ballet 芭蕾鞋   slipper 拖鞋

platform 松糕鞋   insole 鞋垫

bootie 中短靴

over the knee boot 过膝靴   knee high boot 及膝靴   espadrille 帆布鞋

Figure 7-2   Footwear design variation
图7-2   鞋类款式设计

"桑达尔"凉鞋是服装史上最早有文献记录的男女皆可穿的鞋子样式（图7-3）。在16世纪中叶时，又出现了第一款带有小跟的鞋。但那时设计的高跟鞋主要不是用来穿的，而是供别人欣赏的。它象征着穿着者的社会名望和财富。

Figure 7-3　Sandal of ancient Egypt
图7-3　古埃及凉鞋

Since the 19th century a lot of women flooded into the labor market and they went for comfortable and resistant shoes that allowed them to combine working and daily life. Thus, pumps were popular since then. The name "pumps" came from the sound produced when walking on a polished surface.

从19世纪以来，随着大量的女性投入社会工作，她们很想寻求一种舒适耐穿，既适合工作又便于生活的鞋子。于是，（中）高跟鞋从那时起流行起来。"pumps"的名字来自穿着它走在光滑的地板上发出的声音。

For centuries, men and women have shared practically the same type of footwear, and men also wore shoes with a medium heel. Since the 19th century, men preferred boots. At the beginning of the 20th century, the Oxford was popular; in 1930s, the Moccasin appeared: flexibility and comfort are the essential characteristics of this type of shoe. In recent decades, sports footwear has stood out, characterized by its great flexibility, comfort and resistance (Figure 7-4).

几个世纪以来，男性和女性穿的鞋子款式几乎是一样的，男士也穿带有中跟的鞋子。自19世纪以后，男士们偏爱穿靴子；20世纪初期，牛津鞋非常流行；20世纪30年代马克森鞋出现了：宽松舒适是这类鞋子的基本特征。在近几十年，运动鞋以其良好的柔韧性、舒适性和耐穿性脱颖而出（图7-4）。

Further reading: the difference between pumps and stilettos

The difference between them is in heel size.

Pumps have a medium-high or short heel; they are usually closed-toe or peep-toe without any fastening or straps. Pumps can be worn as casual as well as formal shoes by both men and women.

Stilettos can be with straps or covered but would definitely have a thin long heel. The tip diameter of the thin, long and narrow stiletto heel is usually less than 1cm. Stilettos are only worn by women on formal and party occasions and often going with evening dress.

延伸阅读：（中）高跟鞋和细高跟鞋的区别

它们之间的区别在于鞋跟的尺寸。

（中）高跟鞋是中高跟或矮跟，鞋头是全包或露趾的，鞋子没有任何固件或鞋带。（中）高跟鞋既可正式穿也可休闲穿，男女皆可用。

细高跟鞋可以系鞋带或包住脚面，但肯定会有一个细而长的鞋跟，又细又长的鞋跟直径通常在1厘米以内。细高跟鞋只是女性在正式或宴会场合中穿着，通常与晚礼服搭配。

Oxford 牛津鞋

loafer 乐福鞋

moccasin 马克森鞋

lace up boot 系带靴

rain boot 雨靴

moto boot 摩托靴

shell toe 贝壳头鞋

snow boot 雪地靴

high top sneaker 高帮运动鞋

slip on sneaker 易穿式运动鞋

running shoe 跑鞋

sneaker 运动鞋

hiking shoe 登山鞋

Figure 7-4   Footwear design variation
图7-4   鞋类款式设计

7.2.3 Headwear 帽饰

The hat is an accessory that, in its beginning, was designed to protect and cover the head. Nevertheless, the initial practical function has moved to the background and, today, hats have also become a fashionable design element following the new trends closely (Figure 7-5).

帽子作为配饰，最初是用来保护和遮盖头部的。然而，这种最初的实用功能已经退居到其次。现在，帽子也已经成为一种紧跟潮流的时尚设计要素（图7-5）。

The first hat in western history was derived from the medieval helmet and had a protective function, as much physically as psychologically, for the person who wore them. From its beginnings, hats consisted of two parts: the crown and the brim. In the first models, the crown served to show your status: the higher the crown, the higher the social class of its wearer. Nevertheless, and after World War Ⅱ, the hat lost this symbolic value and became a purely decorative accessory.

西方最早的帽子源于中世纪时期的头盔，它给戴用者提供身体上和心理上的防护。一开始帽子是由两部分组成：帽身和帽檐。在最初的样式中，帽身是用来显示身份的，帽身越高，戴用者的社会阶层就越高。但是"二战"以后，帽子便逐渐失去了这种象征意义，而变成了一种装饰性的饰品。

In the first half of the 20th century, hats were a key complement in the female wardrobe. However, from the 1960s it became less common to wear a hat, and its use was relegated to special occasions as just a complementary or ornamental piece. The hat has also been an indispensable part of men's outfits throughout history that can indicate the wearer's social status. From the 1960s the hat lost some of its popularity; but at present it has recovered some presence, for instance, it is often used as a protection against rain, snow and sun at sporting events.

在20世纪上半叶，帽子是女性衣橱中一个重要的配件。但是到20世纪60年代以后戴帽子就变得不那么常见了，帽子成为仅在特定场合里使用的辅助性或装饰性饰物。帽子有史以来也是男装中不可或缺的一部分，能够体现佩戴者的社会地位。但从20世纪60年代以后帽子便失去了人气，不过现在在某些场合又开始使用了，比如在体育赛事中，帽子常用来抵御雨雪和阳光。

homburg 小礼帽

picture hat 阔边女帽

cloche 钟形帽

baseball cap 棒球帽

bowler hat 圆顶礼帽/波乐帽

flat cap 鸭舌帽

turban 头巾帽

wide-brimmed hat 宽边帽

silk hat 高筒礼帽

beret 贝雷帽

sun visor 遮阳帽

party hat 宴会帽

Figure 7-5　Headwear design variation
图7-5　帽子款式设计

7.2.4　Jewelry 首饰

Jewelry is an ornamental object considered indispensable for both men and women, since it helps to enhance and individualize the outfit it accompanies (Figure 7-6). Jewelry such as necklaces, earrings, rings, brooches, hairpins etc. can beautify the wearer's clothing, hair and body. In ancient times, jewelry was often made of precious materials, like gold, silver and jade. It had a symbolic nature, indicating social status, and was attributed with a noticeable religious value. Nevertheless, in the present, some jewelry is designed to be novel and funny; consumers pay more attention to its aesthetic and artistic value. In addition to using traditional precious materials like diamond, platinum, gems and crystal to make jewelry, some inexpensive materials are also frequently used, like acrylic, resin, wood, shells, artificial gems, artificial crystal etc. Jewelry design is not mainly for the purpose of helping wearers show off, but to represent their taste and individual style, or just to add some fun to their lives.

　　首饰是男女都不可缺少的装饰性物品，它可以让服装提升形象，增强个性（图7-6）。项链、耳环、戒指、胸针和发夹等首饰可以美化佩戴者的服装、头发和身体。在古代，首饰经常用一些贵重的材料像金、银和玉制作。它具有象征性，表示社会地位，并具有较强的宗教价值。然而，现代的首饰设计得新颖而有趣味，消费者更看重它的审美和艺术价值。除了用传统的珍贵材料像钻石、铂金、宝石和水晶制作首饰以外，一些廉价的材料也经常使用，如亚克力、树脂、木头、贝壳、人造宝石、人造水晶等。首饰设计的主要目的不是让佩戴者自我炫耀，而是要表现他们的品位和个性化风格，或仅仅给生活带来一些乐趣。

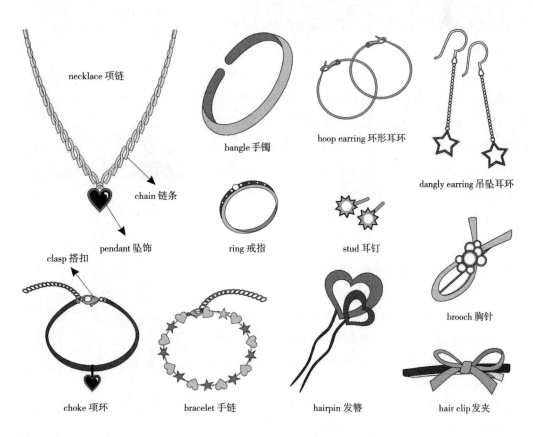

necklace 项链

bangle 手镯

hoop earring 环形耳环

dangly earring 吊坠耳环

chain 链条

pendant 坠饰

ring 戒指

stud 耳钉

brooch 胸针

clasp 搭扣

choke 项环

bracelet 手链

hairpin 发簪

hair clip 发夹

Figure 7-6　Jewelry
图7-6　首饰

7.2.5 Neckwear 颈饰

Neckties and scarves are both elegant neckwear (Figure 7-7). Presented in a wide range of colors, textures and patterns, they can add a touch of distinction or a note of color to any suit or dress. The necktie is a masculine accessory and has nearly become the standard companion to a man's suit. It is a sign of elegance: it enhances the shirt style and emphasizes the verticality of the body. Currently it is not unknown in women's fashion as it often appears in women's neutral style fashion design. As for the scarf, sometimes it is used to keep the neck warm, but most of the time to beautify the face. It can be knotted around the neck or around the shoulders like a cape. If the scarf is large enough, it can also be wrapped into a top, as figure 7-8 shows.

领带和丝巾都是雅致的颈部装饰（图7-7）。颈饰以其丰富多样的颜色、材质和图案可以给套装或连衣裙增添些许不同或一抹色彩。领带是男性化的饰品，几乎已经成为男式西服套装的标配。它是绅士的标志；可以提升衬衫的格调，突出笔挺的身材。现在，因其也经常出现在中性化的女装设计中，所以在女装中使用也并不罕见。丝巾有时是用来给脖颈保暖的，但更多的时候是为了美化面部。它可以围绕在脖子上；或像小披风一样搭在肩部；如果丝巾足够大，还可以缠裹成一件上衣，如图7-8所示。

scarf 丝巾

tie 领带   bowtie 领结

Figure 7-7　Neckwear
图7-7　颈饰

7.2.6 Small accessories 小饰品

Small accessories refer to those articles that accompany garments, ranging from belts, stockings, gloves, to small gadgets for keeping hi-tech items like a cell phone or a tablet. All have a practical function but, increasingly, they show the wearer's individuality and preferences.

小饰品是指那些搭配服装的小物件，从腰带、长

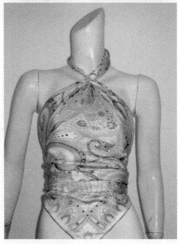

Figure 7-8　Scarf
图7-8　丝巾

筒袜、手套，到盛放手机或平板电脑等高科技物品的小配件。虽然这都是些实用的小配件，但它们也越来越多地展示出佩戴者的个性和喜好。

The belt had its origin in the military world, where it was used as a means of carrying arms. At present, belts have been narrowing or widening according to what fashion dictates. When men's and women's belts are not limited to just holding up clothes, they have taken various forms and assumed a more decorative role. Women usually emphasize their narrow waist by wearing a belt. The sash is a long strip of cloth worn around the waist, which can be made into a flower or bow knot, often used to ornament tops or dresses. Suspenders (BrE braces) are a pair of straps that pass over the shoulders and are fastened to the top of pants at the front and back to hold the pants up (Figure 7-9).

腰带最早出现于军事领域，用于携带武器。现在，随着时尚变化的要求，腰带时窄时宽。当男女的腰带不仅仅局限于固定服装时，便可以产生多种系法，且更具有装饰性。女性佩戴腰带可以突显腰身。装饰性腰带是一根系在腰上的长布条，系成花朵或蝴蝶结状，常用来装饰上衣或裙装。吊裤背带是一对绕过肩部，在裤子前后顶端固定的带子，可以将裤子提起（图7-9）。

Gloves were usually used to cover the hands and keep warm throughout history. However, at present gloves have become decorative accessories as well, and frequently appear on the catwalk. The designs of gloves are diverse: some are very short, that can only cover fingers; some are very long that can cover the whole arm.

手套有史以来就是遮盖手部保暖用的。但是现在手套也已成为一种装饰性配饰，并经常出现在时装舞台上。手套的设计非常多样化：有些很短，只能遮住手指；有些又很长，可以遮盖住整个手臂。

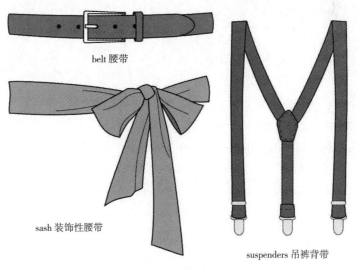

belt 腰带

sash 装饰性腰带

suspenders 吊裤背带

Figure 7-9　Belts
图7-9　腰带

117

Hosiery includes socks, stockings and panty hose (BrE tights); it can be sheer, semi opaque or opaque, and made of nylon, lycra, cotton or wool. Stockings partially cover the legs and sometimes are fastened by a garters belt (BrE suspender belt). Panty hose totally cover the legs and abdomen up to the waist. Stockings were historically a unisex accessory with the legs remaining hidden due to the requirements of strict social norms. Nevertheless, from the 20th century stockings and panty hose were only worn by women, and currently it prevails that women wear panty hose in autumn and winter as well (Figure 7-10).

袜子包括短袜、长筒袜和连裤袜；它可以透明、半透明或不透明，用尼龙、莱卡、棉或羊毛制作而成。长筒袜包裹住大腿的一部分，有时需用吊袜带固定。连裤袜全包住腿部和小腹直到腰部。长筒袜在历史上是男女通用的配饰，由于严格的社会规范的要求，双腿要一直包裹隐蔽。但是从20世纪以后，长筒袜和连裤袜就只被女性穿着了，现在女性在秋冬季节也流行穿连裤袜（图7–10）。

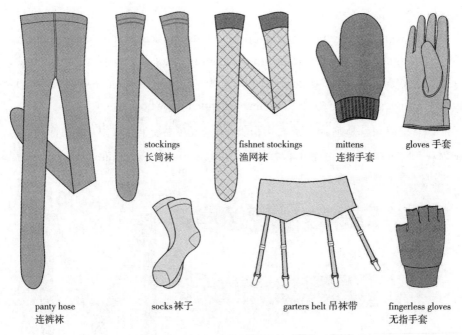

stockings
长筒袜

fishnet stockings
渔网袜

mittens
连指手套

gloves 手套

panty hose
连裤袜

socks 袜子

garters belt 吊袜带

fingerless gloves
无指手套

Figure 7-10　Gloves and hosiery
图7-10　手套和袜子

本章小结

■ 配饰是服装设计中不可缺少的一部分，具有画龙点睛的作用。

■ 现代饰品既具有实用性，更具有装饰性。

■ 包、鞋、帽、首饰、颈饰和小饰品都有多种类型和各自的发展历史。

Exercises 课后练习

1. Please discuss the role of accessories with your partner.

请与你的同伴讨论配饰的作用。

2. Analyze and distinguish each type of accessories' form and features.

分析和辨别各种配饰的造型和特征。

3. Please discuss with your partner how to use accessories completing the outfit.

请与你的同伴讨论如何用配饰来完善服装搭配。

Chapter 8
第八章

Women's Clothing Design
女装设计

课题名称：Women's Clothing Design

课题内容：女装设计特征

女装款式设计

课题时间：2课时

教学目的：让学生学习并掌握各种女装设计风格及主要款式特点的表达。

教学方式：结合PPT与音频等多媒体教学课件，以教师课堂讲解为主，学生随
堂练习与小组讨论为辅。

教学要求：1. 掌握相关词汇。

2. 能够用英语描述出自己设计作品的风格和款式特征。

课前（后）准备：课前预习关于女装设计的基础知识，浏览课文；课后记忆单
词，并参照图片描述各种女装的风格及款式特征。

8.1 The characteristics of the female figure and women's clothing design
女性体型和女装设计的特点

8.1.1　The characteristics of the female figure 女性体型特点

To an adult woman, her standard stature is about seven times as high as her head height. Her shoulders are a bit narrow and slant; the breast and buttocks are plump; the waist is slim, so the body shape is like an hourglass (Figure 8-1). The skin fat is slightly thick, which makes the curves of the body surface smooth. The female figure is graceful and charming; therefore, women's clothing is more beautiful and changeable than men's.

对一个成年女性来说，标准身高差不多是头长的七倍。肩部稍窄且略微倾斜，胸部和臀部丰满，腰部纤细，所以身体形状像一个沙漏（图8-1）。皮下脂肪稍厚，使体表曲线光滑顺畅。女性的体型优美而性感，因此女装比男装更加美观和富于变化。

Figure 8-1　The characteristics of the female figure
图8-1　女性体型特点

8.1.2　The characteristics of women's clothing design 女装设计的特点

Women's clothing design mostly focuses on the variations of styles, the combination of details, and the application of adornments. Women's clothing styles are variable with new styles being constantly created to replace the old ones. Colors of women's clothing are bright and vivid; the patterns and fabrics are abundant to choose from. In every season, plenty of new products are promoted to the market but their prevailing period is short. A lot of designers engage in women's clothing design and this work is highly competitive, but the impetus of fashion development is mainly derived by women's clothing design.

女装设计主要注重款式的变化、细节的组合和装饰的运用。女装款式多变，新款式不断创造出来取代旧款。女装色彩明艳，图案和面料运用丰富。每一季都会有大量的新产品上市，但流行的时间却很短。有许多设计师从事女装设计，使这项工作的市场竞争非常激烈，但时尚发展的动力主要源自女装设计。

8.2 The style of women's clothing design
女装设计的风格

8.2.1 Minimalist style 简约风格

The silhouette and structure of clothing are not very complicated; the adornment and detail designs are scarcely used (Figure 8-2). "Less is more" is its design principle.

服装的廓型和结构精炼，装饰和细节设计尽量简化（图8-2）。"以少胜多"是它的设计原则。

Figure 8-2 Minimalist style (Sinoer)
图8-2 简约风格（新郎－希努尔女装）

8.2.2 Chinese traditional style 中式传统风格

The design inspiration of this style comes from Chinese traditional culture and arts, or Chinese folk arts (Figure 8-3). The mandarin collar, slanting front, edge binding, Chinese frogs and traditional embroideries have become the symbols of Chinese traditional clothing culture. Some auspicious patterns, like peony, lotus and orchid are also its cultural icons. Nowadays Chinese traditional style clothing design is frequently presented in both domestic and international designers' works.

这种风格的设计灵感来自中国传统的文化和艺术，或中国民间艺术（图8-3）。立领、偏襟、绲边、盘扣和传统刺绣已经成为中国传统服饰文化的代表性符号。一些吉祥图案像牡丹、莲花和兰花也是其文化象征。现如今中式传统的服装设计频繁地出现在国内外设计师的作品中。

Figure 8-3 Chinese traditional style (Beijing institute of fashion technology)
图8-3 中式传统风格（北京服装学院新中装发布）

8.2.3 Neutral style 中性风格

This style of clothing obscures the gender difference between male and female, that is, women's clothing has some characteristics of men's clothing. Designers may use some elements of men's clothing in their design. Neutral style clothing is basically unadorned using simple forms and plain colors.

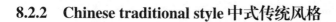

123

这种风格的服装减弱了男女之间的性别差异，也就是说，女装具有男装的一些特征。设计师可能会在设计中运用男装的一些元素；中性风格女装基本上无装饰，形式简洁，色彩朴素。

8.2.4　Casual style 休闲风格

Because casual style clothing is worn comfortably and practically, it is very popular in our modern life. Relative to the clothing worn on formal or working occasions, it has less limitation in design. Therefore, it has plenty of design motifs and forms. Designers can search various inspirations to create these works.

因为休闲风格的服装穿起来舒适且实用，在现代生活中非常流行。相对于在正式或工作场合穿着的服装，它的设计受限较小。因此，它有着丰富的设计主题和形式。设计师可以寻找多方面的灵感来创作作品。

8.2.5　Romantic and extravagant style 浪漫奢华风格

This style of clothing generally refers to the haute couture or evening dresses worn at formal parties or ceremonies. Designers will use delicate fabrics and adornments to design them; and some parts of the clothing also need to be made manually. Ladies will look graceful and captivating when wearing these clothes (Figure 8-4).

这类服装一般是指穿于正式宴会或典礼的高级时装或晚装。设计师会运用精致的面料和饰品进行设计，有些服装的局部还需要专业的手工制作。穿着这类服装的女士优雅而迷人（图8-4）。

8.2.6　Avant-garde style 前卫风格

When designing avant-garde style of clothing, designers often use exaggerated or even strange forms and peculiar materials to express their unique design ideas. Such clothes usually have a strong visual impact so they are very attractive (Figure 8-5). Young people are fond of wearing them since they consider these

Figure 8-4　Romantic and extravagant style (NE·TIGER)
图8-4　浪漫奢华风格（东北虎高级定制礼服）

clothes able to show their individuality.

设计前卫风格的服装时，设计师经常运用夸张甚至怪异的形式和奇特的材料，来表达他们独特的设计理念。这类服装通常具有强烈的视觉冲击力所以非常受人瞩目（图8-5）。年轻人尤其喜欢穿，并以此来展现张扬的个性。

8.3 The categories of women's clothing design
女装设计的分类

Figure 8-5   Avant-garde style
图8-5   前卫风格

8.3.1   Jackets and coats (Figure 8-6) 上衣和外套（图8-6）

Tailored jacket 西装外套

This is a tailored style of women's jacket whose length is around the hip. There are 1-3 buttons on the placket and 2-4 buttons near the sleeve opening. It is fitted at the waist, so it can show women's figure better than some other looser garments.

这是精致裁剪的女装款式。衣长在臀围上下；前襟有1—3颗纽扣；袖口附近有2—4颗纽扣。它腰部合体，所以比起其他宽松的女装，可以更好地展现女性的身材。

Cardigan style jacket 开襟式外套

This style of jacket has a front opening; can be fastened or unfastened. It is a little loose-fitting and usually has no collar. Woven fabrics and knitted fabrics are both suitable to make it. The jacket made of knitted fabrics is more favored by women because of its soft feel.

这种上衣是对襟式，可系合也可敞开。它略微宽松，通常没有领子，梭织面料和针织面料都适合制作这种服装。针织面料做成的开襟式外套因其手感柔软而更受女性青睐。

All-weather coat 全天候外套

An all-weather coat is worn on the very outside of all garments. It has various lengths: above the knee, at the knee and below the knee; the knee-length style is the most popular.

It is characterized by having two wrist straps on the sleeves, two flap pockets on the front, and two epaulets on the shoulders. It is decorated with a self-belt which makes the wearer look slender. The all-weather coat evolved from men's trench coats in the First World War. This coat can be easy to coordinate with outfits according to the temperature in spring and autumn, so it becomes a necessary item in women's wardrobes.

全天候外套穿在所有衣服的最外面。它有多种长度：膝盖以上、齐膝和膝盖以下；齐膝的款式最流行。它的特点是袖子上有腕带；前身有两个带盖的口袋；肩上有肩襻；饰有一条本色腰带，让穿着者更显苗条。全天候外套是由一战时男装中的战壕式风衣演变而来的。在春秋季节，它易于根据气温搭配服装，所以成为女性衣橱中的必备品。

Chanel suit style jacket 夏奈尔套装式上衣

Chanel suit was designed by Coco Chanel and is one of her most famous design works. This style of jacket is collarless, and is characterized by hem-stitching and employing the particular "Chanel tweed". It was designed about 100 years ago, although the fashion trend had changed greatly, it is still alive in the 21st century.

夏奈尔套装是由可可·夏奈尔设计的，是她最著名的设计作品之一。这类上衣没有领子，其特点是镶边和用独特的"夏奈尔"粗花呢制作。它是在一百年前设计的，虽然当时的时尚潮流已经时过境迁，但在21世纪依然流行。

Short jacket 短上衣

A short jacket is a little shorter and the length is around the waist. It can be form-fitting or loose-fitting. If matched appropriately, wearing a short jacket can make a wearer look taller, e.g. matched with long pants or a short skirt.

短上衣稍短一点，长度在腰部上下。它可以合体也可以宽松。如果搭配得当，穿短上衣可以使穿着者看起来更高挑一些，比如搭配长裤或短裙。

Denim jacket 牛仔上衣

A denim jacket is a leisure garment which can be worn on many occasions. Double stitches and metal buttons are its typical characteristics. Although its style design is relatively formulaic, the details have likewise changed a lot following the fashion dynamics through the years.

牛仔上衣是一种可以在很多场合穿着的休闲服。双缉明线和金属纽扣是它的典型特征。虽然它的款式设计相对程式化，但长时间以来，其细节也随着时尚动态发生了很大变化。

tailored jacket 西装外套　　　cardigan style jacket 对襟式外套

denim jacket 牛仔上衣

Chanel suit jacket 夏奈尔套装上衣　　　short jacket 短上衣

Figure 8-6　Jackets and coats
图8-6　上衣和外套

all-weather coat 全天候外套

8.3.2　Blouses 衬衫

Base style blouse 基本款衬衫

It has something in common with a shirt, but has more variations than men's shirt (Figure 8-7).

类似于（男士）衬衫，但比男士衬衫更富有变化（图8-7）。

Peplum style blouse 裙腰式衬衫

A peplum style blouse consists of two parts: a bodice and a peplum which is attached to the waistline (Figure 8-8). The peplum is a flared frill; it can be made of self fabric or another fabric. Peter Pan collar is very common in this style.

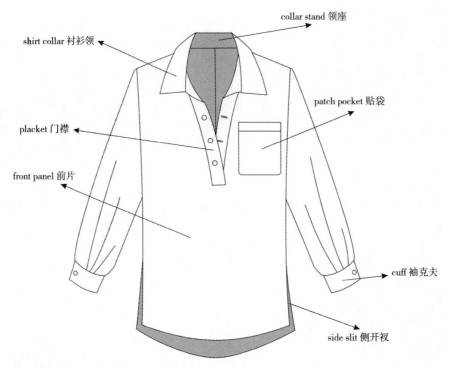

collar stand 领座

shirt collar 衬衫领

patch pocket 贴袋

placket 门襟

front panel 前片

cuff 袖克夫

side slit 侧开衩

Figure 8-7　Base style blouse
图8-7　基本款衬衫

裙腰式衬衫由两部分组成：合体的上衣和衔接在腰部的小短裙（图8-8）。小短裙是一条喇叭状的裙边；可以用本色面料或另一种面料制作。"彼得·潘领"（小圆领）是这种风格的常见领型。

Mock two-piece style blouse 假两件式衬衫

It consists of two parts: the inside shirt and the outside knitwear (Figure 8-9). Actually, the inside shirt is not complete; the invisible part does not exist and the visible part is sewn on the outside knitwear. The knitwear can be a pullover or a cardigan. Looking at the whole, this outfit seems to consist of two pieces, but in fact it is just one piece.

它由两部分组成：里面的衬衫和外面的针织衫（图8-9）。实际上，里面的衬衫是不完整的；看不见的部分并不存在，看得见的部分缝在了外面的针织衫上。针织衫可以是一件套头衫或开衫。从整体上看，这件衣服像是两件套，但实际上它只是一件。

Figure 8-8　Peplum style blouse
图8-8　裙腰式衬衫

Figure 8-9　Mock two-piece style blouse
图8-9　假两件式衬衫

8.3.3　Dresses (Figure 8-10) and skirts (Figure 8-11) 连衣裙（图8–10）和半身裙（图8–11）

Base style dress 基本款连衣裙

The modeling of a base style dress is simple and does not have any redundant decorations; the length is generally above the knee, having or no sleeves. It can be easy to be matched; ladies usually wear it under a coat or a tailored jacket. The silhouette of it can be H-line, A-line or X-line.

基本款连衣裙的造型简洁，没有多余的装饰；长度一般在膝盖以上；有或没有袖子。它很容易搭配，女士们经常穿在外套或西服以内。它的廓型可以是H型、A型或X型。

Cheongsam style dress 旗袍裙

Cheongsam style dress can also have many variations in addition to the traditional classic type. For instance, the slits can be placed on the front, back or no slits at all; the length can be below the knee or above the knee; the neck can be drop-shaped or petal-shaped. Besides silk, it can be made of cotton, linen or lace as well.

旗袍式连衣裙除了传统的经典款还可以有很多变化。例如，开衩可以在前面、后面或没有开衩；长度可以在膝下或膝上；领形可以是水滴形或花瓣形。除了丝绸，还可以用棉布、亚麻或蕾丝制作。

Beach dress 沙滩裙

A beach dress is very long and can almost touch the floor. It is a kind of summer holiday clothing, which is derived from the Bohemia style and often designed with bright colors and large floral patterns. Silk and chiffon are the best choice to make beach dresses, because they can produce a gracefully romantic effect. The beach dress has neither collars nor sleeves, and the halter neck is very common. It can go with a straw hat or a sunhat, flip flops and sunglasses.

沙滩裙非常长，几乎可以垂到地面。它是一种夏季度假服装，源于波西米亚风格，通常用亮丽的颜色和大花图案进行设计。丝绸和雪纺是制作沙滩裙的最佳面料选择，可以产生优美浪漫的效果。沙滩裙无领无袖，吊带领很常见。它可以与草帽或太阳帽、人字拖和太阳镜相搭配。

Blouson dress 宽松束腰连衣裙

Blouson is originally a jacket which draws tight at the waist and is loose above the waistband. It takes most of the traits of the American bomber jacket. A blouson dress is loose at the upper part but close fitting from the waist to the bottom; many office ladies often wear this style of dress.

"布拉桑"原来是一种腰部收紧、腰带以上宽松的夹克。它吸收了美国飞行夹克的大部分特点。宽松束腰连衣裙的上身宽松，腰部到底摆贴体，许多白领丽人经常穿着这种风格的服装。

Trapeze dress 梯形裙

It is a very loose A-line dress that swings away from the body and does not cling to the body. This dress style has become very popular in recent years.

这是一种非常宽松的A字形连衣裙，摆动于身体之上，不贴体。这种裙型近年来非常流行。

Evening dress 晚礼服

An evening dress is worn on formal occasions in the evening, like cocktail parties or celebrations. It is very form-fitting and commonly has a full length; sometimes a dress is tailored for just one certain customer and is finely sewn. Ladies should be more elegant and noble if matched with a fine clutch purse, high heels and exquisite jewelry.

晚礼服穿于晚间正式场合，如鸡尾酒会或庆典仪式。它非常合体，通常长及脚面；有时一件礼服只为一位顾客量身订制，且制作精美。女士们搭配精致的手袋、高跟鞋和精美的首饰后会更加优雅高贵。

Wedding dress/wedding gown 婚纱礼服

A wedding dress is worn by a bride during the wedding ceremony. It is generally in all white symbolizing the virginity of the bride. Some wedding dresses are strapless style, while some have attached sleeves. The lower part of the wedding dress can be a bell skirt or a fishtail skirt, and sometimes has a train. Wedding dresses are mainly made of brocade, gauze or lace. When wearing a wedding dress, the bride also wears a white wedding veil and takes a bouquet in her hands.

　　婚纱礼服是新娘在婚礼时穿着的。它通常是全白色，象征着新娘的纯洁。有的婚纱是抹胸式的，有些附有袖子；婚纱的下半部分可以是钟形裙或鱼尾裙，有时带有拖裙。婚纱主要由锦缎、薄纱或蕾丝制作。穿婚纱时新娘还会戴一条白色的头纱，手里拿着花束。

base style dress(X–line and A–line)
基本款连衣裙（X型和A型）

trapeze dress
梯形裙

cheongsam style dress
旗袍裙

blouson dress
宽松束腰连衣裙

beach dress 沙滩裙

wedding dress/wedding gown 婚纱礼服

evening dress 晚礼服

Figure 8-10　Dresses
图8-10　连衣裙

131

Miniskirt 迷你裙

The hemline of a miniskirt is well above the knee, generally at mid-thigh level or even shorter. A dress with such a hemline is called a miniskirt dress. A miniskirt can be matched with tailored jackets, blouses and T-shirts. It is very popular among young girls because it can show their slender legs; and can also be worn by tennis players, cheerleaders and dancers.

迷你裙的底摆在膝盖以上，通常在大腿的中间位置，甚至更短。有这种底摆长度的连衣裙称作"超短连衣裙"。迷你裙可以与西服、衬衫和T恤相搭配。它在年轻女孩中很受欢迎，因为可以展现她们纤细的双腿；也是网球运动员、啦啦队和舞蹈演员的惯常穿着。

Pegged skirt 楔形裙

A pegged skirt is a little wider at the hip but tapers gradually towards the bottom. Its length is normally below the knee. It sometimes has slits on the front, back or sides for women to walk conveniently.

楔形裙在臀部稍宽，但逐渐向底摆收紧，长度一般在膝盖以下。有时会在前面、后面或侧面开有衩口，以方便女性走路。

Straight skirt 直筒裙

Straight skirts can have various lengths: at the knee, above the knee and below the knee.

直筒裙可以有多种长度：齐膝、膝盖以上和过膝。

Full skirt/bell skirt 蓬蓬裙

A skirt with fullness gathered into the waistband. Its shape is just like an umbrella, a bell or a reversed bud. It can make a woman have a young and lovely look; notice that it is suitable for a slender lady but not for a chubby one.

一种造型饱满并在腰部收紧的裙子。其外形就像雨伞、吊钟或倒立的花苞。它可以使女性拥有年轻可爱的外表，值得注意的是，它适合身材苗条而非丰满的女士。

Flared skirt/A-line skirt 喇叭裙

A flared skirt fits at the waist but flares out to the hemline. Its silhouette is like an A shape or a triangle. It can make women look elegant and modest, and once was very popular in the 1950s.

喇叭裙在腰部合体，但是向下摆张开。它的廓型像一个A字形或三角形。它可以让女性看起来优雅端庄，在20世纪50年代曾经非常流行。

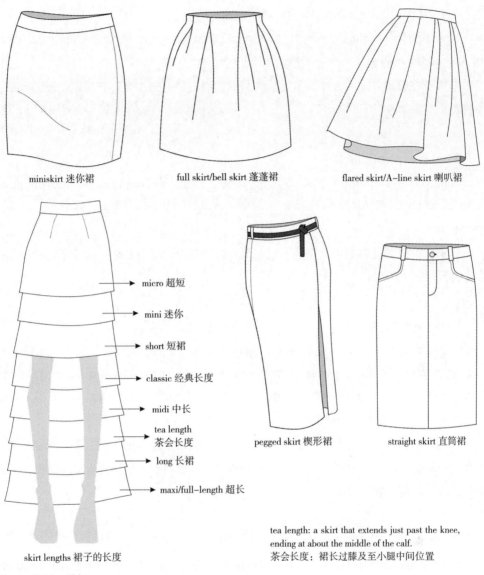

miniskirt 迷你裙

full skirt/bell skirt 蓬蓬裙

flared skirt/A-line skirt 喇叭裙

micro 超短

mini 迷你

short 短裙

classic 经典长度

midi 中长

tea length
茶会长度

long 长裙

maxi/full-length 超长

pegged skirt 楔形裙

straight skirt 直筒裙

skirt lengths 裙子的长度

tea length: a skirt that extends just past the knee,
ending at about the middle of the calf.
茶会长度：裙长过膝及至小腿中间位置

Figure 8-11　Skirts
图8-11　半身裙

8.3.4　Pants (Figure 8–12) 裤子（图8–12）

Hot pants 热裤

Hot pants are very short and tight with the waistband below the navel, and the leg openings are around the joint of the buttocks and thighs. Women who have slim legs are suitable to wear this style of pants.

热裤非常短也非常紧身，裤腰在肚脐以下，裤口差不多是在臀部与腿部的衔接处。双腿修长的女性适合穿着这种类型的裤子。

Flared pants/bellbottoms 喇叭裤

Flared pants are form-fitting at the hip and then gradually become wide from above the knee to the bottom; they can make the leg lines more graceful.

喇叭裤在臀围处非常合体，然后从膝盖以上到裤脚逐渐加宽；它可以使腿部线条更加优美。

Tapered pants 锥形裤

Tapered pants are form-fitting or a little loose at the hip, but become narrower at the bottom; the shape is just like an awl. They can make the wearer's legs look long, straight and slim.

锥形裤在臀部合体或稍微宽松一点，但在裤脚变窄，造型就像一个锥子。它可以使穿着者的双腿显得修长、挺直和纤细。

Leggings 打底裤

Leggings are very skinny, usually in black. They are worn under a wool jumper or a wool skirt in winter; and are worn under a long blouse or a miniskirt in spring and autumn. The length can be at the mid-calf or near the ankle. Leggings are often made of fabrics which are blended with spandex fibers, so they are worn very comfortably.

打底裤非常紧身，通常是黑色。冬天穿在毛衫裙或羊毛裙的里面，春秋季穿在长衫或短裙的里面。其长度可以在小腿中部或脚踝附近。打底裤一般用含有氨纶纤维的面料制作，所以穿起来非常舒服。

Capri pants 卡布里裤（合体七分裤）

Capri pants are a close-fitting type of pants ending between the knee and the ankle. "Capri" is named after the Isle of Capri in Italy where these pants were first worn.

"卡布里裤"（合体七分裤）是一种长度在膝盖和脚踝之间的紧身裤子。"卡布里"是以意大利的卡布里岛命名的，在那里这种裤子最早被穿用。

Cropped pants 宽松七分裤

The length of cropped pants is similar to that of Capri pants, and can end anywhere on the calf. The difference between Capri pants and cropped pants is that the former is formfitting but the latter is looser. Cropped pants became very popular in the United States in the 1950s; they were known as clam diggers at that time. The shorter length of cropped pants makes them ideal for bicycle riding, as long pant legs may become tangled in the bike chain.

宽松七分裤的长度和卡布里裤相似，可以长及小腿的任何位置。卡布里裤和宽松七分裤的区别在于前者合体，而后者较宽松。宽松七分裤在20世纪50年代的美国非常流行，在那时它被称为"挖蛤裤"。宽松七分裤因为长度稍短非常便于骑自行车，而长裤的裤腿则有可能会被自行车的链条缠住。

Culottes 裙裤

Culottes were once a type of men's clothing in European history, but nowadays they usually refer to women's wear. They look like a skirt, however they are actually pants. They also come down to between the knee and the ankle.

"克尤罗特"（裙裤）在欧洲历史上曾经是一种男装，但是现在通常指女装。它看起来像裙子，但实际上是裤子。其长度在膝盖和脚踝之间。

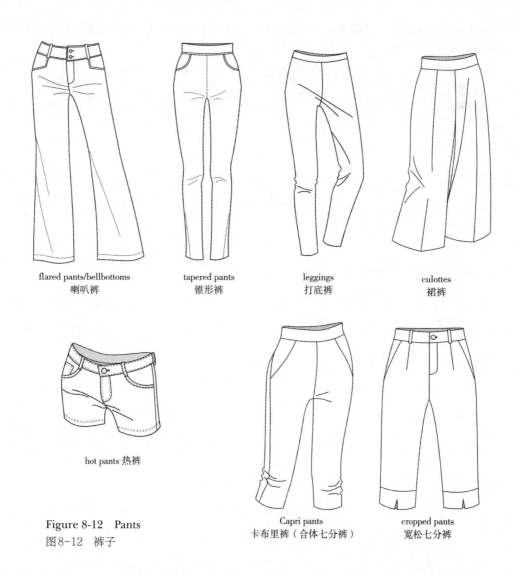

flared pants/bellbottoms
喇叭裤

tapered pants
锥形裤

leggings
打底裤

culottes
裙裤

hot pants 热裤

Figure 8-12　Pants
图8-12　裤子

Capri pants
卡布里裤（合体七分裤）

cropped pants
宽松七分裤

本章小结

■ 女性身材曲线玲珑，女装比男装设计效果更美观，且富有变化性。

■ 女装设计有多种风格，每种风格有其不同的设计特点。

■ 女装款式多种多样，每种款式皆有其丰富的细节特征。

课后练习

1. Please retell the characteristics of each clothing design style and garment mentioned above.

请复述以上每种服装设计风格和各款服装的特点。

2. Please search for some pictures of women's clothing and describe the design.

请搜寻一些女装的图片并描述其设计。

3. Group exercise: role play 小组练习：角色扮演

Suppose you are a designer, Please explain your design works to other students as if they were your clients or bosses.

假设你是一名设计师，其他同学是客户或老板，请向他们介绍你的设计作品。

Chapter 9
第九章

Men's Clothing Design
男装设计

课题名称： Men's Clothing Design

课题内容： 男装设计特征

男装款式设计

课题时间： 2课时

教学目的： 让学生学习并掌握各种男装设计风格及主要款式特点的表达。

教学方式： 结合PPT与音频等多媒体教学课件，以教师课堂讲解为主，学生随

堂练习与小组讨论为辅。

教学要求： 1. 掌握相关词汇。

2. 能够用英语描述出自己设计作品的风格和款式特征。

课前（后）准备： 课前预习关于男装设计的基础知识，浏览课文；课后记忆单

词，并参照图片描述各种男装的风格及款式特征。

9.1 The characteristics of the male figure and men's clothing design
男性体型和男装设计的特点

9.1.1 The characteristics of the male figure 男性体型的特点

Most adult men are about 1.7-1.85 meters tall. Men's shoulders are wider than women's but their hips are narrower. Their chest is broad and the torso is thick, so the upper body is strong. Men's body shape is like an inverted trapezoid or an inverted triangle (Figure 9-1). Men are taller, stronger than women, and full of strength.

多数成年男性的身高在1.7—1.85米。男性的肩部比女性宽，但胯部略窄。他们胸部宽阔，躯干厚实，上半身比较发达。男性的体型像倒梯形或倒三角形（图9-1）。男性比女性身材高大、健壮，且充满力量。

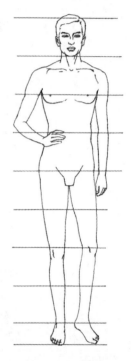

Figure 9-1 The characteristics of the male figure
图9-1 男性体型的特点

9.1.2 The characteristics of men's clothing design 男装设计的特点

Because of the restriction of the male figure and their perceived role as needing to be socially responsible, men's clothing is relatively stable and changes subtly compared with women's. The feature of men's clothing is that the silhouette is not greatly changed; more attention is paid to the design of details, using good quality fabrics and fine tailoring to make. There are fewer decorations on men's clothing, and the significance is mainly to show a person's social status and economic ability. Most colors of men's clothing are subdued; black, grey, dark blue and khaki are frequently used to show men's modesty. Additionally, men generally value the brand of a garment when they choose it; therefore, for the promotion of men's clothing products, the influence of a brand plays a vital role.

受男性体型和所承担社会职责的限制，与女装相比，男装的变化相对稳定而微妙。男装的特点是廓型不会有很大的变动，更注重细节设计，并运用优质的面料和精致的裁剪制作。男装的装饰较少，其意义主要在于体现一个人的社会地位和经济能力。多数男装的颜色都很沉着，黑色、灰色、深蓝和卡其经常使用，以表现男性的稳重。此外，男士们在选择服装时通常很看重它的品牌，因此在男装产品推广中，品牌的影响力起到举足轻重的作用。

9.2 The style of men's clothing design
男装设计的风格

9.2.1　Classic style 经典风格

The classic style of clothing generally refers to the suit (Figure 9-2). Such a suit derived from the 19th century men's tuxedo, which was conservative, orthodox, formulaic, and had some consistently matching approaches. However, with the development of history and society, the style of tuxedo had been simplified. A standard suit includes a tailored jacket, a shirt, pants and a tie. The suit colors are usually black or purplish blue. Men are used to wearing this style of suit on business or formal occasions.

经典风格的男装通常指西服套装（图9-2）。这种套装源自19世纪男士的塔式多礼服，它是保守的、正统的、程式化的，并有一些固定的搭配方式。然而随着历史和社会的发展，塔式多礼服被

Figure 9-2　Classic style (Sinoer)
图9-2　经典风格（新郎－希努尔男装）

简化了。标准的西服套装包括外套、衬衫、裤子和领带，套装的颜色一般为黑色或藏蓝色。男士们习惯在商务或正式场合穿着这种套装。

9.2.2　Business style 商务风格

This type of clothing is a variation of the suit jacket, but it can be a little more casual, and has various styles. The collar of the jacket can be lapels or a stand collar; there are 3-5 buttons or a zipper closure on the placket, flap pockets or patch pockets on the front. In terms of color, warm colors can also be used, such as khaki or beige. This style of clothing is also suitable for wearing on official or business occasions.

这种类型的服装是正式西服的变化款，可以更休闲化，且有多种样式。上衣的领子可以是翻领或立领，前襟有3—5粒纽扣或拉链，前身有插袋或贴袋。在颜色方面，也可以使用暖色，如卡其或米色。这种风格的服装适用于公务或商务场合。

9.2.3　Casual style 休闲风格

Casual style clothing is worn during leisure time instead of working hours. Jackets, jeans, shirts and T-shirts are the main categories of this style. Stripes, checks, letters and figures are the frequently used decorative patterns of these garments (Figure 9-3).

休闲风格的服装穿于闲暇时光而非工作时间。夹克、牛仔服、休闲衬衫和T恤是其主要的服装种类。条纹、方格、字母和数字是这些服装常用的装饰图案（图9-3）。

Figure 9-3　Casual style
图9-3　休闲风格

9.2.4　Sports style 运动风格

Because clothing of sports style is worn comfortably and conveniently, it is popular among not only young people but also the elderly. With more and more young people being keen on outdoor sports, the outdoor sportswear also becomes popular, for example, sweatshirts, safari jackets and anoraks.

运动风格的服装穿着舒适方便，无论在年轻人还是老年人中都很流行。随着越来越多的年轻人热衷于户外运动，户外运动装也变得流行起来，如运动衫、猎装夹克和防风衣。

9.2.5　Chinese traditional style 中式风格

Men's clothing of Chinese traditional style

Figure 9-4　Chinese traditional style (2014 APEC male leaders' jacket)
图9-4　中式风格（2014年APEC会议男性领导人服装）

mainly refers to the Tang suit, or variations of the student uniform of the Republican period or the Sun Yat-sen uniform (Figure 9-4). A mandarin collar is their common feature. This type of clothing is often decorated with some Chinese traditional patterns, such as a dragon, auspicious clouds and Chinese calligraphy. The Tang suit is also decorated with Chinese frogs and edge binding.

传统风格的男装主要包括唐装，或民国学生装、中山装的变化款（图9-4），立领是它们共同点。这类服装经常饰有中国的传统文化元素图案，如龙纹、祥云和书法。唐装还饰有盘扣和缇边。

9.2.6　Hip–hop style 嘻哈风格

Hip-hop is a subculture which originated from America. Its music and dance are very popular among young people. Hip-hop style clothing is characterized by a large T-shirt decorated with fancy patterns or doodles, and very loose jeans or harem pants; it is also matched with sneakers and a cap. These young people also like wearing hooded sweatshirts or casual shirts, and prefer to wear some weird accessories (Figure 9-5).

嘻哈是源自于美国的亚文化。它的音乐和舞蹈在年轻人中非常流行。嘻哈风格的服装以饰有趣味图案或涂鸦的大T恤、非常宽松的牛仔裤或哈伦裤为特点，配有运动鞋和棒球帽。这些年轻人还喜欢穿连帽的运动衫或休闲衬衫，喜欢佩戴一些奇特的配饰（图9-5）。

Figure 9-5　Hip–hop style
图9-5　嘻哈风格

9.3 The categories of men's clothing design 男装设计的分类

9.3.1　Jackets and tops 上衣

Suit jacket 套装西服

A suit jacket can be single-breasted or double-breasted; the single-breasted panels always have front cuts. Generally, there are 3-4 pockets on the front panel: one chest welt pocket and 2-3 flap pockets beneath. The chest welt pocket is used to insert a bright-colored handkerchief. There are one or two vents on the back. Suit jackets are always made of fine fabrics, e.g., worsted wool (Figure 9-6).

套装西服可以是单排扣或双排扣，单排扣的前襟为圆底摆。前身上一般有3—4个口袋：一个胸袋，下面2—3个带盖口袋，胸袋用来装色彩鲜艳的胸巾。后身有一或两个开衩。套装西服通常使用精制的面料制作，如精纺羊毛（图9-6）。

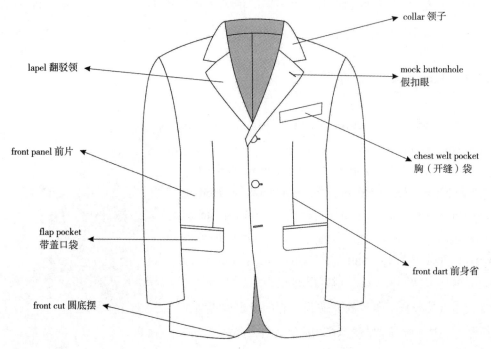

collar 领子
lapel 翻驳领
mock buttonhole
假扣眼
front panel 前片
chest welt pocket
胸（开缝）袋
flap pocket
带盖口袋
front dart 前身省
front cut 圆底摆

Figure 9-6　Suit jacket
图9-6　套装西服

The lapel can be notched or peaked. There is a mock buttonhole on the left lapel, which was once useful but nowadays is hardly used. In the 19th century when a gentleman wanted to please his lover, he often wore a flower in chest, so the buttonhole was used to insert the flower.

翻领可以是平驳领或戗驳领。在左边的领面上有一个假扣眼，它曾经是有用的，但现在几乎不用了。在19世纪当一个绅士想要取悦他的爱人时，胸前往往揣一朵花，这个扣眼就是用来插花的。

Tuxedo shirt 塔士多衬衫
Tuxedo shirt is a formal type matched with men's formal dresses, e.g., the tailcoat or morning coat. It has a wing collar, and is decorated with ruffles, pleats or U-shaped stiff liner on the front panel.

塔士多衬衫是正装衬衫，和男士的正式礼服相搭配，如燕尾服或晨礼服。它有翼领，前身饰有花边、褶裥或U形硬衬。

Vest (BrE waistcoat) 马甲

It is a sleeveless upper garment, usually worn over a shirt and under a formal dress.

马甲是无袖上衣，穿在衬衫的外面，西服的里面。

Blazer 布雷泽外套（休闲西装）

The blazer is similar to the suit jacket and also has lapels. It is single-breasted or double-breasted and often has patch pockets. The obvious characteristic is that it is usually made of navy blue fabrics and decorated with gold or silver buttons. The pants which are matched with a blazer can be changeable. The jacket of some association uniform or school uniform is often designed as a blazer style with a shield-shaped badge attached on the left chest.

布雷泽外套和套装西服相似，也有翻领。它是单排扣或双排扣，通常有贴袋。其明显的特征是用海军蓝色面料制作，饰有金色或银色的纽扣。和布雷泽外套搭配的裤子可以是不同色面料。许多社团制服或校服的上装常设计成布雷泽式，左前胸带有一个盾形的徽章。

Casual jacket 休闲夹克

A casual jacket is loose but tightens at the bottom. It is constituted of many cut pieces and often decorated with stitches. It has a zipper or snap buttons for fastenings; a stand collar or turn-down collar. There are many pockets on the front panel: patch pockets, zipper pockets or welt pockets; each of them is practical. The casual jacket can be made of cotton, denim, man-made fiber fabrics or leather.

休闲夹克衣身宽松，但底部收紧。它由多个裁片构成，经常饰有明线。它有拉链或按扣，立领或翻领。前身有很多口袋：贴袋、拉链口袋或贴边袋，每个口袋都很实用。休闲夹克可以用棉、牛仔、人造纤维面料或皮革制作。

Safari jacket 猎装夹克

A safari jacket is a garment originally designed for the purpose of going on safari in the African bush. When matched with pants or shorts, it then becomes a safari suit. The safari jacket is commonly made of lightweight cotton drill or poplin. It is traditionally in khaki, with a self-belt and two epaulets. It has four or more bellows pockets to make it more practical.

猎装夹克最初是设计用来在非洲丛林中狩猎时穿的。与长裤或短裤搭配，便变成了狩猎套装。猎装夹克一般是用轻质的斜纹棉布或府绸制作。它传统上是卡其色，并带有本色腰带和肩襻。它有四个或更多的风箱口袋，使其更加实用。

Sweatshirt 运动衫

A sweatshirt is a warm piece of clothing for the upper part of the body, with long sleeves, often made of knitted cotton. It is an informal type of clothing; men usually wear it when they do sports or go travelling (Figure 9-7).

运动衫是保暖的上身服装，长袖，通常用针织棉面料制作。它是一种便装，男士们一般在运动或外出旅游时穿着（图9-7）。

blazer 布雷泽外套   tuxedo shirt 塔士多衬衫   vest 马甲

casual jacket 休闲夹克   safari jacket 猎装夹克   sweatshirt 套头运动衫

Figure 9-7　Jackets and tops
图9-7　上衣

9.3.2　Coats (Figure 9-8) 外套大衣（图9-8）

Trench coat 战壕式风衣

This coat evolved from the European military uniform which was used in the trenches during the First World War. The traditional style is double-breasted with 10 buttons, having lapels, a storm flap and two flap pockets with buttons on it. It has a belt at the waist and two buckled straps around the wrists. There are two epaulettes on the shoulders which once had military practicability. Its length is various ranging from just above the knee to the ankle. Trench coat is made of waterproof durable cotton twill fabrics. Although its traditional color was khaki, newer versions of it come in many colors.

这种风衣是从第一次世界大战时在战壕中使用的欧洲军服演变而来。传统的款式是有10颗纽扣的双排扣式；有翻领、披胸布和两个钉有纽扣的带盖插袋；腰部系有腰带，腕部有带扣襻的腕带。肩上有肩襻，肩襻曾经具有军事实用性。衣长有多种样式，从膝盖以上到脚踝不等。战壕式风衣用防水耐用的斜纹棉布制成。风衣传统的颜色是卡其色，而新版的风衣中有多种颜色。

Chesterfield 柴斯特菲尔德外套

A chesterfield is a long, tailored overcoat. Its name originated from a British earl's name in the 19th century; and was a staple of smartly dressed men's wardrobes from the 1920s to 1960s. It can be single or double breasted and has a velvet collar and a fly front. The chesterfield can be made with many kinds of heavy fabrics in different patterns; however, the classic style of it is made of black or charcoal tweed. It is very versatile, that can be worn over casual wear as well as a suit or semi-formal dress.

柴斯特菲尔德外套是长款合体的大衣。它的名字来自19世纪一个英国伯爵的名字，在20世纪20年代到60年代，它是衣着精致的男士们衣橱里的必备行头。它单双排扣都可以，有带天鹅绒的领子和暗门襟。柴斯特菲尔德外套可以用多种不同图案的厚质面料制作，但它的经典款是用黑色或灰色的粗花呢制作的。它的用途非常广泛，既可以穿在西装或半正式服装的外面，也可以穿在休闲装外面。

Pea coat 海员短大衣

A pea coat is generally made of navy-colored heavy wool. It was originally worn by sailors of the European and later by the American navy. It is characterized by short length, a double-breasted placket, broad lapels, large buttons, and vertical or slash pockets.

海员短大衣通常用海军蓝色的厚重毛料制作。它最初是由欧洲水手穿着而后被美国海军穿着。它以短衣身，宽领面，双排扣衣襟，大纽扣，竖式或斜式的插袋为特点。

Duffel coat 达夫尔外套

A duffel coat is a coat made of duffel which is a coarse, thick woolen fabric. The name derives from Duffel, a town in Belgium where the fabric originated from. It was once worn by the British Royal Navy. The duffle coat is notably characterized by an oversized hood and several toggles with loops to attach them to. The hood offers enough room to wear over a navy cap; and the wooden or plastic toggle fastenings are easily to be fastened and unfastened for sailors while wearing gloves in cold weather at sea.

达夫尔外套是用一种粗糙、厚实的羊毛制品达夫尔面料制作而成的。这个名字来自比利时的一个小镇达夫尔，那里是这种面料起源的地方。它曾被英国皇家海军穿着过。达夫尔外套的显著特征是有个超大的风帽和数个带圆环的拴扣。风帽有足够大的空间可以戴在海军帽上面，木制或塑料制的拴扣易于士兵在天气寒冷的海面上戴着手套扣上或解开。

Balmacaan 巴尔玛外套

A balmacaan is a loose, full overcoat with raglan sleeves. It is named after an estate near Inverness, Scotland. It is single breasted, and has fly front and slash pockets. Tweed and gabardine are the common fabrics to make the balmacaan.

巴尔玛外套是一件相对宽松，有插肩袖的大衣。它是以苏格兰因弗尼斯（英国苏格兰北部的一个城市）附近的一个庄园命名的。它是单排扣，有暗门襟和斜插袋。制作巴尔玛外套常用的面料是粗花呢和华达呢。

Anorak 防风衣（连帽风衣）

An anorak is a hooded short coat, sometimes having a cord at the waist or the bottom hem. It has many pockets, which are very practical. It is generally made of functional fabrics which are wind resistant, waterproof and breathable. Men often wear anoraks when doing outdoor exercise or going climbing.

防风衣是件连帽的短外套，有时在腰部或底摆有一根绳带。它有很多口袋，都非常实用。它通常是用防风、防水、透气的功能性面料制作而成的。男士们经常在进行户外运动或登山时穿着防风衣。

trench coat 战壕式风衣　　chesterfield 柴斯特外套　　balmacaan 巴尔玛外套

pea coat 海员短大衣　　anorak 防风衣（连帽风衣）　　duffel coat 达夫尔外套

Figure 9-8　Coats
图9-8　外套

9.3.3　Pants (Figure 9-9) 裤子（图9-9）

Suit pants 西裤

Suit pants are originally worn matching with suit jackets, but now they can also go with shirts or T-shirts. They are more formal than other types of men's pants. They have two side pockets and two rear welt pockets. On the legs of the pants there are two center creases down to the bottom that make the men's legs look straight and long.

西裤原本是和套装西服搭配的，但现在也可以和衬衫或T恤衫搭配。它比其他类型的男裤要正式一些。它有两个侧口袋和两个后开袋。裤腿上有两条裤中线，让男士的腿看起来又直又长。

Cargo pants 工装裤

Cargo pants are a type of loose pants that have many pockets in various places, for example on the side of the leg above the knee. They are originally designed for tough outwork and the fabrics used to make them are usually wear-resistant. If the length is shorter, then they are known as cargo shorts.

工装裤是一种宽松型的裤子，在很多地方都有口袋，如在膝盖上方裤腿的侧面。它最初是为艰苦的户外工作设计的，制作它的面料通常是耐磨的。如果裤长较短，则被称为"工装短裤"。

Bermuda shorts 百慕大短裤

Bermuda shorts can be worn as semi-casual or business attire. The hem, which can be cuffed or uncuffed, is around 1 inch (2.54cm) above the knee. They are available in a variety of colors, including many pastel shades as well as darker shades.

百慕大短裤可穿作半休闲或商务服装。裤口可以卷边或不卷边，其长度大约在膝盖以上1英寸（2.54厘米）处。它在颜色上有多种选择，既有深色调，也包括很多浅色调。

They are so-named because of their popularity in Bermuda, a British overseas territory, where they are considered as appropriate business attire for men. They are made of suit-like fabrics and worn with knee-length socks, a dress shirt, a tie, and a blazer. They are more appropriate to be worn in hot subtropical and tropical climates than the typical heavier clothing favored in Europe. Even nowadays they are also worn on some business occasions in the hotter seasons, especially in the United States, Australia, and New Zealand.

它之所以得名，是因为它在英国海外领地百慕大地区非常流行，在那里它被看作是男士适合的商务着装。它用类似西服的面料制作，搭配齐膝的袜子、礼服衬衫、领带和布雷泽外套。比起在欧洲流行的典型的厚重服装，它更适合在炎热的亚热带和热带气候里穿着。即使现在，在炎热的季节里，它仍穿于一些商务场合，特别是在美国、澳大利亚和新西兰。

Harem pants 哈伦裤

Harem pants were once women's clothing, but nowadays they have become popular with men as well. Harem pants were historically baggy, long, and caught in at the ankle. However, currently they mainly refer to those pants which are loose and have many folds at the drapes, and tightened at the bottom. Besides these, pants with a very low crotch are also regarded as harem pants.

哈伦裤曾经是女性穿的裤子，但是现如今也已成为了男性的流行服装。历史上的

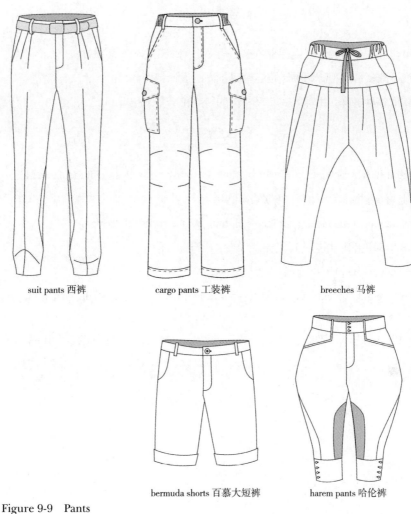

suit pants 西裤　　　cargo pants 工装裤　　　breeches 马裤

bermuda shorts 百慕大短裤　　　harem pants 哈伦裤

Figure 9-9　Pants
图9-9　裤子

哈伦裤又肥又长，并且在脚踝处束住。而现在的哈伦裤主要指那些宽松的，垂下很多褶皱，且裤脚收紧的裤子；另外，裆部很低的裤子也被认为是哈伦裤。

Breeches 马裤

Breeches refer to the knee-length garments and are fastened by buttons, or straps and buckles at the bottom. They were worn by men from the late 16th century to the early 19th century. But they only survived in England as very formal attire since then. Nowadays they are sometimes used for some athletic wear, such as riding breeches and fencing breeches.

马裤指的是齐膝的裤子，在裤口用纽扣或带襻扣住。男士们从16世纪末到19世纪初一直穿马裤。但是从那以后，它只作为非常正式的着装在英国流传下来。现在它有时被用作一些体育运动装，如马术马裤和剑术马裤。

本章小结

- 男性身材高大、健壮，呈倒梯形或倒三角形。男装变化稳定而微妙。
- 男装设计有多种风格，每种风格各有其不同的设计特点。
- 男装款式多种多样，每种款式有其各自的特点和穿着文化。

课后练习

1. Please retell the characteristics of each clothing design style and garment mentioned above.

请复述以上每种服装设计风格和各款服装的特点。

2. Please search for some pictures of men's clothing and describe the design.

请搜寻一些男装的图片并描述其设计。

3. Group practice: role play 小组练习：角色扮演

Suppose you are a designer. Please explain your design works to other students as if they were your clients or bosses.

假设你是一名设计师，其他同学是你的客户或老板，请向他们介绍你的设计作品。

Chapter 10
第十章

Children's Clothing Design
童装设计

课题名称： Children's Clothing Design

课题内容： 童装设计特征

童装款式设计

课题时间： 2课时

教学目的： 使学生学习并掌握儿童不同成长阶段的服装款式特征及穿用性能。

教学方式： 结合PPT与音频等多媒体教学课件，以教师课堂讲解为主，学生随
堂练习与小组讨论为辅。

教学要求： 1. 掌握相关词汇。

2. 能够用英语描述出自己设计作品的风格和款式特征。

课前（后）准备： 课前预习关于童装设计的基础知识，浏览课文；课后记忆单
词，并参照图片描述各种童装款式的特征。

10.1 Characteristics of children's clothing design
童装设计的特点

In the field of garments, childhood refers to the age span ranging from birth to 16 years old. As children grow and develop fast, they will have different requirements for garments at different ages. Therefore, some characteristics call for special attention in children's clothing design.

在服装领域，一般将0—16岁统称为儿童时期。儿童生长发育速度较快，不同的年龄段对服装有不同的需求。因此，在进行童装设计时有些特点需要特别注意。

10.1.1 Safety 安全性

Children's skin is very delicate, especially that of infants and young children, which puts higher requirements on the safety performance of fabrics of children's clothing, including color fastness (water resistance, sweat resistance and abrasion resistance), pH and formaldehyde content. If the color fastness is unqualified, skin injury could occur with the dye being transferred from clothing to skin. The pH, either too high or too low, would expose the skin to bacteria. Besides, excessive levels of formaldehyde may cause respiratory inflammation and skin inflammation through respiratory system or through the contact with skin.

儿童皮肤娇嫩，尤其是婴幼儿，这对服装面料的安全性能提出了更高的要求，包括色牢度（耐水、耐汗、耐摩擦）、pH和甲醛含量等。色牢度不合格，在穿着过程中染料会从服装转移到皮肤上造成伤害。pH过高或过低，都会使皮肤受到细菌的侵害。过量的甲醛则会通过呼吸系统或皮肤接触而引发呼吸道和皮肤炎症。

10.1.2 Comfort 舒适性

In terms of the characteristics of children's skin and body, it would be better to choose fabrics and accessories of natural fiber which is comfortable, breathable, environmentally friendly and moisture-retentive. Besides, fabrics used for children's clothing should have such properties as fine elasticity, flexibility and wrinkle resistance.

根据儿童皮肤和身体的特点，面辅料宜选用舒适环保、吸湿透气的天然纤维。此外，童装面料还要具备良好的伸缩性、柔韧性和抗皱性等性能。

10.1.3　Functionality 功能性

Tight clothing will cause oppression to the body of children, hindering physical movement. In accordance with children's physiological features and active personality, enough garment ease should be left in the design to ensure adequate space for children's physical movement (Figure 10-1).

　　紧身服装会对儿童的身体造成压迫，影响肢体活动。根据儿童的生理特征和活泼好动的特点，在设计时要预留足够的松量，以便给儿童的身体活动提供充足的空间（图10-1）。

Figure 10-1　Functionality for children's clothing
图10-1　童装的功能性

10.1.4　Periodicity 阶段性

In order to design the clothing that can meet the needs of children's growth, designers ought to understand the characteristics of children at different growing periods, including their physiological and psychological features. Meanwhile, both practical and aesthetic functions of the clothing should be realized so as to improve children's aesthetic senses through designer's unique vision and children-attracting elements in the design, and accordingly promote children's mental and physical development.

　　设计师应了解儿童不同成长阶段的特征，包括生理特征和心理特征，以设计出适应儿童成长需要的服装。同时还要体现服装的实用功能与审美功能，用独特的视角和充满童趣的设计提升儿童对美的认识，以促进儿童身心健康的发展。

10.2 Clothing design for children at different ages 不同年龄的童装设计

Children show different characteristics at different ages. According to different growing stages, the whole growing process of children can be divided into namely infancy, early childhood, preschool stage, school-age stage and juvenile.

儿童的成长阶段有很大的差异性。根据不同的成长阶段可分为婴儿期、幼儿期、学龄前期、学龄期和少年期。

10.2.1　Infancy (0-1 years old) 婴儿期（0—1岁）

An infant is characterized by a large head, a short neck, a protruding abdomen as well as short legs bending to the inside; the chest circumference of an infant is almost the same as its waist circumference or hip circumference. Infants also have small gender differences.

婴儿的体型特点是头大，颈短，腹部突出，腿短且向内呈弧状弯曲；胸、腰、臀基本等宽。性别差异小。

Infants' clothing is simple and comfortable with few seams, few structural lines, collarless structure, pastel color, moderately looser cuffs and leg openings. Jumpsuits are more suitable for an infant because they are easy to be put on and taken off, and can better protect the infant's waist and abdomen parts. Instead of hard materials like buttons and zippers, strips of cloth are often used for connection to protect infants' skin. Infants sweat and excrete frequently, so it would be better to choose pure cotton knitted fabrics that are breathable and hygroscopic. Additionally, accessories such as a hat, hand coverings, a small bib and socks can be designed to match the clothing.

婴儿服装的特点是款式简单舒适，接缝线和结构线较少，无领，色彩柔和，袖口和裤口适当宽松。连体衣更适合婴儿，因为它易于穿脱，能更好地保护婴儿的腰腹部。服装通常用布条系接，避免用纽扣及拉链等硬质材料损伤婴儿的肌肤。婴儿排汗排泄较多，所以面料最好选择透气、吸湿性好的纯棉针织物。另外，还可以设计帽子、手套、围嘴、袜子等与服装配套。

10.2.2　Early childhood stage (1-3 years old) 幼儿期（1—3岁）

A young child is always found to have a big head, narrow shoulders, short limbs and a big tummy, thus the body shape resembles a cylinder. Young children grow rapidly during this period and get to take part in more activities than before, so their clothes should be loose, comfortable, and convenient for their movement. In addition, young children feel great affections for patterns and color, especially for the former. The patterns which are funny and colorful are children's favorite ones.

幼儿的体型特征一般是头大、肩窄、四肢短、肚子大，形状像一个直筒。这一时期的幼儿成长迅速，活动增加，服装要以宽松舒适，便于活动为主。另外，幼儿对图案和色彩都很敏感，特别是图案。趣味性强、色彩丰富的图案是孩子们喜欢的。

10.2.3　Preschool stage (4-6 years old) 学龄前期（4—6岁）

Children grow higher and slender during this period. Their shoulders become broad, limbs grow rapidly, and gender differences in clothing become obvious. Thus, ornaments such as lace and drape can be used in girls' garments to highlight their sweetness and loveliness. Boys' clothing design often uses straight-shaped style, and the garment fabrics for them should be durable, wearable, easy to wash and easy to dry.

这一时期儿童的体型开始变得瘦长，肩部增宽，四肢增长较快，服装的性别差异也变得明显。女童的服装可以运用花边、褶皱等装饰来表现女孩的甜美可爱。男童的服装设计多采用直筒型，其服装面料应结实耐穿，易洗快干。

10.2.4　School age (7-12 years old) 学龄期（7—12岁）

There is a rapid increase in children's height during this period, and their bodies become better proportioned, with their heads smaller and necks slenderer in proportion to other parts of their bodies. Gender differences begin to appear between boys' and girls' body shapes.

这一时期儿童身高增长迅速，体型变得匀称，头部比例变小，颈部细长。男童和女童的体型开始出现性别差异。

Since the amount of exercise is increased for children at this stage, the joint parts on clothes such as elbows and knees are easy to get frayed. It is suitable to choose thickened fabrics or big

Figure 10-2　School age
图10-2　学龄期

embroideries and appliqué to make the clothing more wearable and attractive. Girls' clothing design may consider an A-line or X-line silhouette to highlight their femininity, while for boys' clothing design, simple and practical jackets and jeans are the more common styles (Figure 10-2).

由于这一年龄段的儿童运动量加大，袖肘、膝盖等关节部位容易磨损，在设计时可适当加厚或者选用贴绣图案，来起到加固和装饰的作用。设计女童服装时可运用A型或

X型的廓型来凸显女孩的柔美乖巧，而简洁实用的夹克服和牛仔裤则是男童服装设计常用的款式（图10-2）。

10.2.5 Juvenile period (13-16 years old) 青少年期（13—16岁）

In this period, the adolescents' figures are almost fully developed. Boys and girls look obviously different in body shapes; the bust line, waist line and hip line of girls are clearly discernible. Boys' and girls' secondary sexual characteristics become apparent.

这一时期的青少年体型逐渐发育完善。男生和女生的身体外观有明显的差异，女生的胸围线、腰围线和臀围线清晰可辨。男生和女生的第二性征变得明显。

Adolescents tend to possess their own unique personality and autonomy and are keen on pursuing faddish things. For instance, girls usually enjoy popular music, cartoons and idol dramas, while boys are fond of sports and computer games. Therefore, designers should acquaint their preferences and life styles. For adolescents, casual wear and comfortable sportswear which are simple, casual and full of distinctive individuality are their favorite ones (Figure 10-3).

Figure 10-3　Juvenile period
图10-3　青少年期

青少年往往具有自己独特的个性和自主性，并且喜欢追求流行事物。比如，女生喜欢流行音乐、动漫或偶像剧，男生热衷于体育运动或电子游戏。因此，设计师要了解他们的喜好和生活方式。对青少年来说，简洁、随意、彰显个性的休闲装和舒适的运动装是他们的最爱（图10-3）。

Last but not least, to design children's clothing well, designers should understand the desire of children's parents as well as the children's physiological and psychological characteristics. If designers could design and beautify the image of children with parental love and affections, the clothing they design will be beautiful, comfortable and able to display the innocence of children. Nowadays, as children's apparel corporations pay more attention to personality, fashion and brand of clothing, it is put forward higher designing requirements on children's clothing.

最后同样重要的是，要设计出好的童装，设计师除了要了解儿童的生理和心理特征，还要了解他们父母的意愿。用父母般的慈爱和情感去设计、美化儿童形象，就能设计出既美观舒适，又能表现儿童天真无邪的童装。如今，随着童装公司越来越重视服装的个性、时尚性和品牌等方面，对童装设计也提出了更高的要求。

10.3 The categories of children's clothing design 童装设计的分类

10.3.1　Jumpsuits and suits (Figure 10-4) 连体衣和套装（图10–4）

Rompers 婴儿连体衣（哈衣、爬服）

Rompers are a common type of infants' clothing, easy to be put on and taken off, simple in design, convenient for infants' movements, and also convenient for parents to change diapers for the infants. Fabrics like pure cotton and fine flannel are often used for their softness and air permeability.

婴儿连体衣是婴儿常见的一种服装，它穿脱方便，造型简单，便于婴儿活动，也便于父母给婴儿更换尿布。在面料上多使用柔软、透气性好的纯棉布或细绒布制作。

Bib overalls 背带裤

Bib overalls are suitable for 2-3 year-old children. They can not only protect young children from coldness, but also feel loose and comfortable. The length of pants can be adjusted with buttons on the straps, which is very practical for children who grow quickly in this period. Some appliqué can be designed on the bib as aesthetic enlightenment for children.

背带裤适用于2—3岁的幼儿，既可以护体防寒，又宽松舒适。背带上的纽扣可以调节裤子的长短，这对于生长快速的儿童来说非常实用。围兜上可以设计一些贴布图案，会对儿童起到审美启蒙的作用。

Cowboy suit 牛仔装

Warm and wearable cowboy suits are the common choice for children during outdoor activities in spring and autumn. The denim used for children's clothing is usually soft and comfortable after washing and enzyme treatment. Appliqué, embroidery and prints are often used to adorn cowboy suits, and the subjects are always about kids' favorite cartoon images. In addition, some other elements are also common in children's cowboy suits, such as top-stitching, distressing, and raw edges and so on.

保暖且耐穿的牛仔装是春秋时节儿童户外休闲常穿的服装。童装牛仔面料通常都经过水洗和酶处理，比较柔软舒适。贴花、绣花和印花经常用来装饰牛仔装，其设计主题往往是孩子们喜欢的卡通形象。此外，缉明线、做旧处理和毛边等元素也常用于儿童牛仔装中。

School uniform 校服

In China, the traditional school uniforms are main track suits which are looser, and are suitable for children whose bodies are in a rapidly growing period. The color scheme tends to be plain and conservative. In the recent years, some western-style school uniforms have become popular, especially in private schools. Boys' western-style uniforms are pant suits, while girls' are skirt suits. Purplish blue, red and gray are the commonly used colors for these uniforms. The school badge is often embroidered on the left chest part of the uniform.

在中国，传统的校服主要是宽松的运动服，适合身体处于生长发育期的儿童。色彩搭配较为朴素和保守。近些年，一些西式校服很受欢迎，尤其是在私立学校。男孩的西式校服是裤套装，女孩的是裙套装；藏青色、红色和灰色是这些校服常用的颜色；上衣的左前胸还绣有学校的徽章。

rompers 婴儿连体衣

bib overalls 背带裤

cowboy suit 牛仔装

school uniform 校服

Figure 10-4　Jumpsuits and suits
图10-4　连体衣和套装

10.3.2 Tops (Figure 10-5) 上衣（图10–5）

Hoody 卫衣（连帽衫）

A hoody is a jacket or a sweatshirt with a hood. There are pullover hoodies and cardigan hoodies. Hoodies are often designed with a big patch pocket on the front known as the "kangaroo pocket", or have letters or cartoons printed on the chest. They are often made of cotton fabrics, either heavy or light, and the cuffs and the bottom hem are made of stretchy rib fabrics.

卫衣是一种带帽的上衣或运动衫，分套头衫和开衫两种。前身通常有被称为"袋鼠兜"的大贴袋，或者前胸印有字母或卡通图案。它一般用棉织物制作，或厚或薄，袖口及底边常采用有弹性的罗纹织物制作。

Tank top (BrE vest) 背心、汗衫

A tank top is a collarless and sleeveless top, suitable for children engaging in outdoor activities or used as loungewear. Decorative patterns or pockets can be designed to make it lovelier.

背心或汗衫是一种无领无袖的上衣，适合儿童户外运动时穿着或用作家居服。上面可以设计些装饰图案或小口袋，以增加童趣。

Button shoulder sweater 肩纽毛衫

This is a pullover type with buttons on one or both shoulders, which is easier for children to put on. There may be 2-3 buttons on one shoulder, or on a raglan opening which slopes form the neck to the armhole. The buttons are both useful and decorative. On the chest of the sweater designers can design some funny patterns, like Disney animal cartoons, or use some bright color blocks to attract kids.

这是在单肩或双肩上都有纽扣的套头衫，对孩子们来说更容易穿着。单肩上有2—3颗纽扣，纽扣也可能钉在从领口斜向袖窿的插肩袖的开口上。除了实用性，这些纽扣还具有装饰性。在毛衫前胸，设计师可以设计些如迪士尼卡通动物等图案，或者装饰一些鲜艳的色块来吸引孩子们。

Hooded vest 连帽马甲

A Hooded vest is used to keep out the wind and cold and always worn over a shirt or sweater. It's convenient to protect the children from temperature changes when they go outdoors. Fabrics adopted are mostly denim, corduroy, knitted fabrics, quilted-down fabrics and so on.

连帽马甲用来抵御风寒，穿在衬衫或针织衫的外面。它便于孩子们在外出时根据气温来搭配服装。其制作面料以斜纹棉布、灯芯绒、针织面料和羽绒、夹棉等居多。

hoody
卫衣（连帽衫）

tank top (BrE vest)
背心、汗衫

button shoulder sweater
肩纽毛衫

hooded vest
连帽马甲

Figure 10-5　Tops
图10-5　上衣

10.3.3　Dresses (Figure 10-6) 连衣裙（图10-6）

Sundress 背心裙

A sundress is a loose, sleeveless dress that does not cover the arms, neck or shoulders. Girls wear it in hot weather. The lower part of it can be redesigned into a straight skirt, a full skirt, a pleated skirt and so on. If the shoulder parts of a sundress are fastened with straps, the length of the dress can be adjustable; this is suitable for little girls who keep growing taller.

背心裙是一件宽松、无袖的连衣裙，将手臂、颈、肩裸露在外。它是女孩们在热天里穿着的。下面的裙子可以再设计成直筒裙、蓬蓬裙或百褶裙等。如果背心裙的肩部是用带子系接，则裙子的长度是可以调节的，这适用于正在长身高的小女孩们。

Swing dress 摆裙

The swing dress which has a wide hemline may make a little girl look more lively and lovely. The loose dress body has many drapes, and its shape varies as the number of seams or size of the dress hemline changes. A wide hemline is fit for children whose abdomens are protruding, and helps them move freely.

摆裙的下摆宽大，使小女孩更显活泼可爱。宽松的裙身有很多自然垂褶，其造型随着裙片数量和裙摆大小的变化而变化。宽松的裙摆适合腹部突出的儿童，而且让她们的身体活动自如。

Princess dress 公主裙

A princess dress gets the waist part nipped, dividing the dress into a formfitting upper part and an A-line skirt with the hemline near the knee; sometimes the waistline seam is raised a little bit. The decorative details can be diversified; bows, flounces, frills and embroidery are the common ornamental techniques. For color selection, pink is the most favored.

公主裙的腰部收紧，并将服装分为合体的上半身和A字型的下裙，裙子的底摆在膝盖上下，有时腰节线还会稍微提高一点。装饰细节可以多种多样，蝴蝶结、荷叶边、皱褶边和刺绣都是常见的装饰手法。在颜色的使用上，粉色是最受欢迎的。

sundress 背心裙　　　　swing dress 摆裙　　　　princess dress 公主裙

Figure 10-6　Dresses
图10-6　连衣裙

10.3.4　Bottoms (Figure 10-7) 下装（图10-7）

Tutu skirt 芭蕾舞裙（图图裙）

A tutu skirt evolves from a female dancer's costume worn in a ballet performance.

Nowadays it mainly refers to a type of little girls' short skirts consisting of many layers of stiff material. The tutu skirt is frilly and somewhat stiff, sticking out from the waist and hip. It is often made of organza, gauze, tulle and silk, and the light color like white, pink and pastel blue is common for tutu skirts.

芭蕾舞裙是从芭蕾舞女演员的服装中演变而来的。现如今主要指由多层硬质材料组成的小女孩穿的短裙。芭蕾舞裙带有褶边，稍稍硬挺，并从腰臀部向外展开。它常用欧根纱、薄纱、网眼纱和丝织品制作，白色、粉色和浅蓝色等浅色是常用的颜色。

Joggers 慢跑裤

Joggers are a kind of comfortable and lightweight sport pants, suitable for children engaging in physical exercise, and often worn with hoodies or T-shirts. The waist part is adjustable with a drawstring or an elastic band. The leg openings can be tightened or loosened through a rib structure, elastic bands, hook-and-loops or zippers. Besides, the pockets of joggers have various types, such as welt pockets, patch pockets, and zip pockets and so on.

慢跑裤是一款适于儿童活动的运动裤，舒适轻便，常搭配卫衣或T恤穿着。裤腰采用抽绳或松紧带调节，裤口可以用罗纹、松紧带、魔术贴或拉链等方法调节松紧。慢跑裤口袋的类型多种多样，有挖袋、贴袋和拉链口袋等。

Jersey shorts 运动短裤

Jersey shorts originally go with a jersey but sometimes can also be worn separately. They are often made of fine soft cotton fabrics with good flexibility and air permeability, so they are usually worn in summer by boys. Like joggers their waistband can be adjustable as well. They can be designed to have or no edge binding, be in solid color or be decorated with characters. The more common decorative method is to add bright color stripes on the sides of shorts legs.

 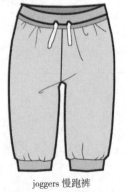 

tutu skirt 芭蕾舞裙　　　　joggers 慢跑裤　　　　jersey short 运动短裤

Figure 10-7　Bottoms
图10-7　下装

运动短裤原本是和针织运动衫搭配穿着的，但有时也可单穿。它一般用柔韧性和透气性好的柔软棉织物做成，所以男孩们在夏季经常穿着。和慢跑裤一样，其腰带也可以调节；可以设计成有或者没有镶边；素色的或饰有字母图案的。最常用的装饰方法是在裤腿的两侧加上亮丽的色条。

本章小结

- 童装设计不仅要考虑美观度，更要考虑安全性与舒适性。
- 童装要根据不同年龄段特征进行设计，每个阶段有其不同的设计特点。
- 童装款式多种多样，每种款式皆有其丰富的细节特征。

课后练习

1. Please search some clothing pictures of children at different ages and describe their designs.

请搜寻一些不同年龄段的童装图片并描述它们的设计。

2. Group task: role play 小组练习：角色扮演

Suppose you are a designer and other students are your clients or bosses, please explain your design to them.

假设你是一名设计师，其他同学是你的客户或老板，请向他们介绍你的设计作品。

Chapter 11
第十一章

Fashion Drawing
服装效果图

课题名称：Fashion Drawing

课题内容：服装效果图

课题时间：2课时

教学目的：让学生学习并掌握服装效果图的绘画工具、材料和绘画技法，以及电脑图像的种类、特点和绘图软件应用的相关术语。

教学方式：结合PPT与音频等多媒体教学课件，以教师课堂讲解为主，学生随堂练习与小组讨论为辅。

教学要求：1. 掌握相关词汇。

2. 能够用英语简单描述出手绘服装效果图的技法和绘画过程，以及电脑图像的种类、特征及软件的运用。

课前（后）准备：课前浏览课文，准备绘画材料和工具；课后记忆单词，做手绘和电脑绘图的练习并描述绘画过程。

11.1 Hand drawing
手绘效果图

11.1.1 Sketching a fashion drawing 起草服装效果图

When some ideas or inspirations emerge in a designer's mind, they need to be expressed through visual media. Many designers will start by drawing a fashion drawing, rendering their ideas on a fashion figure (model) to get a feel for what the garments will look like on the human form. Being able to draw fashion drawings is one of the basic skills for a designer to master.

当设计师的头脑中产生了设计想法或灵感以后，需要通过某种媒介表现出来。很多设计师会从画服装效果图开始，把想法描绘在时装人体（模特）身上，以感受服装穿在人身上的样子。画服装效果图是设计师需要掌握的基本技能之一。

Before drawing the fashion drawing, designers need to know something about the human bodies' framework on which a garment will hang. There are obvious differences in the male and female forms; however, they both can be broken down into the same simple block shapes. The head can be portrayed as an egg; the chest as a wastepaper basket; the pelvic area as a gymnastic vaulting horse; the limbs as tapering tubes; the feet and hands as cones; and the joints as small balls (Figure 11-1). Once we have drawn these block shapes in the relative size, we notice that the sketch is like a wooden dummy (Figure 11-2). We can move some of these shapes around just like moving some parts of the dummy's body to create the pose we want.

在画服装效果图之前，设计师有必要了解一下支撑服装的人体框架结构。虽然男女的体型明显不同，但它们可以被分解成相同的简单块状。脑袋可以描绘成一个鸡蛋，上半身像一个纸篓，骨盆处像体操鞍马，四肢像尖端的管子，手脚像锥体，关节像小球（图11-1）。一旦我们把这些形状以相应的比例画出来，会发现所画的草图像一个木头人（图11-2）。我们可以移动其中的一些形状，就如同转动木头人的肢体，以塑造我们想要的姿势。

Many are of the opinion that designers should take anatomy classes to help improve their sketching. Anatomy provides them with a better understanding of the structure and mechanics of the body, from which they can gain a greater appreciation for how clothing will fit and flow around the human form.

许多人认为设计师应该学一些解剖学来帮助提高绘画能力。解剖学使他们更好地理解人体结构和力学，对服装是如何适应和围绕人体，他们从中可以获得更好的鉴别力。

ASHION DESIGN ENGLISH
服装设计英语

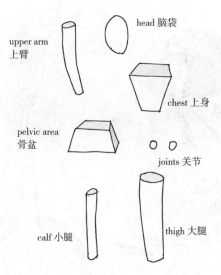

Figure 11-1　Block shapes of human body
图11-1　人体的块状

Figure 11-2　Wooden dummy
图11-2　木头人

There is a convention in fashion drawing that a figure should be elongated to give it more elegance. However, that elongation should involve only the legs. A real-life body will measure around seven-and-a-half-head sizes, but in fashion drawing this increases to nine heads, with the extra length added below the waist. All the other proportions of the basic block shapes should be based on reality. There is another thing that should be kept in mind, that the depicting figures are alive and their poses are lively.

画服装效果图有一个惯例，就是将人物画得瘦长以增添美感。通常是将腿部加长。真人的身材差不多是七个半头高，但在效果图中要增加到九个头高，额外的长度都增加到腰部以下。其他所有的身体比例应基于现实。另外还有一点需要记住，所描绘的人物是有生命的，其姿态要生动活泼。

When the framework is decided, next is to draw the flesh coverage of the figure, and depict its face, hands and shoes. Then it is to draw the clothes; this is the most important step to draw a fashion drawing. When drawing the clothes, designers should consider the structure of the body and garments, and how the garments drape over the body, so as to make the drawing look correct (Figure 11-3). The strokes of contours have different density. When drawing those garment parts which cling to the body, the lines will be thick; while for the opposite, drawing detached parts, the lines will be thin. In order to improve the skill of drawing, designers can often do some practice sketches. It is advisable to complete these within a limited time, such as 15 or 20 minutes. You need not worry about the final outcome of the picture, but just draw what you want boldly. This is a good practice method which can help designers improve the ability of responding quickly and drawing skillfully.

upper arm 上臂　head 脑袋　chest 上身　pelvic area 骨盆　joints 关节　calf 小腿　thigh 大腿

人物框架确定下来以后，下面就是画附着的肌肉；刻画面部、手和鞋。然后是画衣服，这是画效果图最重要的一步。画衣服时，设计师需要考虑人体和服装的结构，以及服装是如何"穿在"人身上的，以使绘图看起来准确（图11-3）。

Figure 11-3　Hand drawing (Painter: Li Xinyao)
图11-3　手绘效果图（绘画：李昕遥）

轮廓线要浓淡适宜。在画贴近人体的服装部分时，线条要重些；相反，画离体的部分时，线条要浅些。为了提高绘画技巧，设计师可以经常画些速写。速写最好是在有限的时间内完成，如15或20分钟。不必担心最终的画面效果如何，只要大胆地表现你想表现的。这是一种很好的练习方式，有助于设计师提高快速反应能力和熟练的绘画能力。

11.1.2　Rendering a fashion drawing 绘制服装效果图

To sketch the fashion drawing, designers only use a pencil and an eraser; while when rendering the fashion drawing, they will use many other materials. As figure 11-4 shows, these are some tools and material for drawing.

起草服装效果图只需用到铅笔和橡皮，而绘制效果图则会用到很多其他材料。如图11-4所示，是一些绘画工具和材料。

Gouache is the most common pigment for fashion drawing. It has the features of both oil paint and watercolor paint. We can use gouache along with brushes (thick or medium) or paintbrushes (medium or thin) to color the garments; and use the thin brush to draw the contours. Thick paintbrushes are suitable to paint the background of the fashion drawing. Notice should be paid to not fill up the colors on the garments, but leave some areas unfilled to simulate the effect of being lit by a distant light. Designers also need to figure out how shadows will fall on the body, the face, and the garments. Selecting a position for the light source will make it easier to be consistent with the placement of shadows.

水粉是画效果图最常用的颜料。它兼具油画和水彩颜料的特点。可以用水粉结合毛笔（大号或中号）或方头画笔（中号或小号）给服装上色；用小号毛笔勾勒轮廓线。大

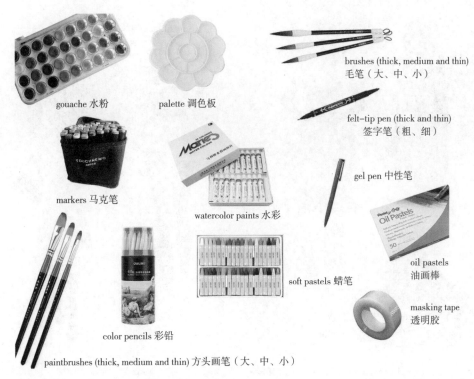

gouache 水粉

palette 调色板

brushes (thick, medium and thin)
毛笔（大、中、小）

felt–tip pen (thick and thin)
签字笔（粗、细）

markers 马克笔

gel pen 中性笔

watercolor paints 水彩

oil pastels
油画棒

soft pastels 蜡笔

masking tape
透明胶

color pencils 彩铅

paintbrushes (thick, medium and thin) 方头画笔（大、中、小）

Figure 11-4　Tools and material for drawing
图11-4　绘画工具和材料

号的方头画笔适合画效果图的背景。注意不要将服装的颜色涂满，留出一些空白，以模拟从远处受光的效果。设计师还需要弄清楚阴影是如何投射在身上、脸上及服装上的；选择一个光源的位置易于使阴影的方向保持一致。

Designers can also try to combine gouache with other materials, such as watercolor paint, color pencils, and oil pastels (Figure 11-5). Unusual combinations of techniques lend a freshness and originality to fashion drawings to make them stand out from the crowd.

设计师还可以尝试将水粉和其他材料结合使用，像水彩、彩铅和油画棒（图11–5）。不同寻常的技法组合会给效果图增添一种新鲜感和独创性，从而使它们脱颖而出。

Besides having a good grasp of drawing tools and materials, designers should also learn how to depict the texture of textiles as it relates to the effect of the drawing. For instance, when rendering a sheer fabric, designers can paint a filmy wash of color layered over skin tone and any other fabrics it rests on. Lace work is also layered over other colors, but demands strong bias cross-hatching to create the illusion of netting on which the pattern of lace is built. Feather and fur require layers of color to create depth at the center and a wispy lightness along the perimeter. Knits are depicted by building up ribs, cables, or twists. In short, experimenting with various techniques is essential for a designer to improve his or her drawing ability.

Figure 11-5　Hand drawing (Painter：Hao Jiawen）
图11-5　手绘效果图（绘画：郝佳雯）

除了掌握绘画工具和材料以外，设计师还应学习如何刻画纺织品的材质，因为它关系到时装绘画的效果。例如，在表现透明面料时，设计师可以在皮肤色和其他面料上面用湿画法画上薄薄的一层。蕾丝的画法也是覆盖在其他颜色上面，但需要画出明显的斜向交叉线，以塑造出蕾丝图案形成的网状错觉。羽毛和皮草需要颜色分层，以塑造出中央色深厚重，周围色浅纤盈的效果。针织物是通过创建罗纹、捻线或拧绳来刻画的。总之，尝试各种技法对设计师提高绘画能力而言是必不可少的。

11.2 Computer-aided drawing
电脑辅助绘画

11.2.1　Computer graphics and drawing software 电脑图像和绘图软件

In addition to traditional ways of drawing, designers can create digital fashion drawings using computer art software (Figure 11-6). Common design software includes Photoshop, CorelDraw, Illustrator, Painter, Sai, and Freehand and so on. All of the techniques described for fashion drawing created by hand can be recreated using the on-screen tools, layers, brushes, and filters. The advantage of digital art is that it is cleaner and paperless, and it allows the designer to update, correct, and reproduce visual information more easily and quickly. Each piece of software has its own specialty and advantages, so designers should choose the appropriate ones and aim to use them proficiently. However, none of these programs is a substitute for drawing by hand. After all, a computer and software are just tools like a brush and a pencil.

除了传统的绘画方式，设计师还可以运用电脑艺术软件创造数码效果图（图11-6）。常用的设计软件包括Photoshop，CorelDraw，Illustrator，Painter，Sai，Freehand等。所有前面描述的手绘效果图的技法都可以运用屏幕上的工具、图层、笔刷和滤镜进行再创作。数码艺术的优点是更干净，无纸化，使设计师更容易和快捷地更新、修改和复制视觉信息。每款软件都有它的特点和优点，所以设计师可以选择合适的软件并熟练运用它们。但是，这些程序中没有一种可以代替手绘，毕竟，电脑和软件就像画笔和铅笔一样仅仅是工具。

Figure 11-6　Computer-aided drawing (Painter: Yan Wenqian)
图11-6　电脑绘画（绘画：闫文茜）

Before choosing the design software we should know something about computer graphics. There are two types of computer graphics: raster graphics (bitmaps) and vector graphics. Raster graphics consist of a great deal of pixels; every pixel has its own color and location information. When zooming in on a raster image, it will be blurred and jagged, and appear with very many pixels; when zooming out, it turns to be clear and photo-realistic again. The resolution determines the crisp quality of an image. Raster graphics are characterized by rich color effects. Photoshop, Sai and Painter are the editors to deal with this type of graphics.

在选择设计软件之前应先了解一下电脑图像。电脑图像有两种类型：位图和矢量图。位图是由大量的像素点组成，每个像素点都有它的颜色和位置信息。当一个位图图像放大时，会变得模糊和锯齿状，并出现很多像素点，缩小后就又变得清楚逼真了。分辨率决定了图像的清晰质量。位图以色彩丰富为特征。Photoshop，Sai 和 Painter 是处理这一类图像的编辑器。

Vector graphics have nothing to do with the pixels and resolution. They are created mathematically with geometric formulas. Different computing methods define the different outlines and color fills of graphics. Vector graphics will be printed very crisply even when enlarged, but they do not have as much color information as raster graphics. CorelDraw, Illustrator, Freehand are all vector graphic editors; they are suitable to create graphic design, page layout, logos, cartoons, etc.

矢量图与像素、分辨率无关。它们是用几何公式以数学计算的方法创建出来的。不同的计算方法决定了不同的图像轮廓和色彩填充。矢量图在放大以后也可以清晰地打印出来，但是它没有位图那样丰富的色彩信息。CorelDraw，Illustrator，Freehand 都是矢量图形的编辑器；它们适合制作图形设计、页面布局、标志、卡通形象等。

Here we choose Photoshop and CorelDraw as the main tools to introduce. Each editor has its advantages and limitations, but many of their functions can complement each other. A good understanding of each operating technique can help designers work efficiently.

这里我们选择 Photoshop 和 CorelDraw 作为主要工具进行介绍。每款编辑器都有它的优点和局限性，但很多功能是相得益彰的，深入了解每一种技法可以帮助设计师高效率地工作。

Photoshop is developed and published by Adobe systems. It has a good capability of image processing so that it can create many fantastic artistic effects. It can simulate various strokes of real brushes, and save plenty of pattern and texture material. Fashion designers can use it to color design sketches, add textures, create a three-dimensional effect, and add background to the fashion drawing. CorelDraw is developed by Corel Corporation. It is more practical in creating complicated curves and is easy to adjust the shapes of graphics. So it is suitable for drawing working sketches of fashion design to show the structure clearly.

Photoshop 是由（美国）奥多比公司研发和发行的。它有着强大的图像处理性能，可以创造出很多神奇的艺术效果。它可以模拟各种画笔的笔触，储存大量的图案和材质素材。设计师可以用它给线稿图上色，添加材质，塑造立体效果，以及给效果图添加背景。CorelDraw是由（加拿大）科立尔公司研发的。它在绘制复杂曲线时更加实用，并易于调整图形形状。所以它适合绘制服装的款式图，使结构表现得清楚准确。

11.2.2　The methods of computer–aided drawing　电脑绘画方法

Designers can either partially or completely use the computer software to create a fashion drawing. If partially using the computer, designers should draw sketches on paper first. It is better to highlight the stroke of lines and keep the background clean. And then scan the sketch or take a photo of it to input into the computer.

设计师可以部分地或完全地运用电脑软件来创作服装效果图。如果是部分地运用电脑，设计师需要先在纸上画好草图。最好笔触画重一些，且保持纸面干净。然后扫描草图或拍照输入电脑。

Open Photoshop, and then color the sketch. This step is mainly to determine the palette and tone of the design. Next step is to draw the details, adding shades and tints to the colors and make the fashion drawing look more three-dimensional. If required, add some textures and patterns to the drawing clothing to enrich its design effects. This is the most time-consuming step including rendering the figure's face, hands, shoes and accessories (Figure 11-7).

打开Photoshop，对草图进行上色。这一步主要是确定色彩搭配和色调。然后是刻画细节，给颜色增添明暗，让效果图看起来更加立体。如果有需要，就给服装添加上材质和图案，以丰富设计效果。这一步是最耗时的，其中还包括绘制人物的面部、手、鞋和饰品（图11-7）。

At last paint the background of the fashion drawing, e.g. add a shadow to the figure or draw some special patterns to foil the figure; some design descriptions can also be typed on the background.

最后是绘制效果图的背景，比如给人物添加投影，或者画一些独特的图案以烘托人物，还可以写上设计说明。

If designers completely use the computer to draw the fashion drawing, they can start with rendering the figure's face. Lay down colors of different shades and tints, and paint with different tips and size of brushes to make the face more photo-realistic. Then render the skin. These works can be done in Photoshop.

如果设计师完全用电脑绘制效果图，可以从刻画人物的面部开始。铺上不同明暗的颜色，运用不同笔尖、不同大小的画笔进行绘画，让面部看起来更加逼真。然后画皮肤。这些工作可以用Photoshop完成。

Figure 11-7　Steps of computer–aided drawing
(Painter: Hao Jiawen/Meng Wenhui/Dong Xiaowen)
图11-7　电脑绘画步骤（绘画：郝佳雯、孟文惠、董晓文）

Next draw the figure's pose and clothing. Designers can use the "pen tool" in Photoshop or "freehand tool" in CorelDraw to draw the curves. Make sure the figure's pose is proper and the clothing proportion is correct. And then render the clothing, shoes, accessories and also the background; the steps are similar to the former method. Sometimes designers can use some Photoshop filter commands to render the drawing, which can produce fine artistic effects.

接下来是画人物的姿势和服装。设计师可以用Photoshop中的"钢笔工具"或CorelDraw中的"手绘工具"绘画线条。要确保人物姿势恰当，服装比例准确。然后是刻画服装、鞋、饰品，还有背景，步骤和前一方法类似。有时设计师可以运用Photoshop中的滤镜命令来渲染效果图，会产生很好的艺术效果。

CorelDraw is more suitable for drawing working sketches (Figure 11-8) and detail sketches. A working sketch refers to the front

Figure 11-8　Working sketches
（Painter:Yue Zichen）
图11-8　服装款式图（绘画：岳姿辰）

view or back view of a garment, which can show the structure and outline clearly. A detail sketch is to break down or zoom in on some parts of the garment, which can further illustrate or highlight the fashion design details. When drawing them, firstly use the "freehand tool" in CorelDraw to draw a rough garment shape; then use the "shape tool" to adjust some nodes on the garment outlines, till to be perfect.

CorelDraw更适合绘制服装的款式图（图11-8）和分解图。款式图是指服装的前视图或背面图，可以清晰地表现服装的结构和轮廓。分解图是拆分或放大服装的某些局部，可以进一步说明或强调设计细节。画这两种图时，首先用CorelDraw中的"手绘工具"画出服装的大致形状，再用"形状工具"调整服装轮廓上的节点，直到造型完美。

本章小结

- 服装效果图是表达设计的媒介，绘制服装效果图分为手绘和电脑绘画。
- 手绘服装效果图分为起草和上色两大步骤。
- 电脑图像分为位图和矢量图。
- 不同的绘图软件有不同的特点和功能，可以结合使用。

Exercises 课后练习

1. Draw a fashion drawing to express your design ideas, and tell your classmates which materials and techniques you have used.

画一张服装效果图来表达你的设计想法，并告诉同学们你所使用的绘画材料和技法。

2. Use Photoshop to paint a fashion drawing; use CorelDraw to draw a working sketch. And interpret the drawing processes to your classmates.

用Photoshop画一张效果图，用CorelDraw画一张款式图。向同学们阐述你的绘画过程。

Chapter 12
第十二章

Developing a Line
服装产品系列研发

课题名称： Developing a Line

课题内容： 服装产品系列策划

服装产品系列研发

课题时间： 2课时

教学目的： 让学生学习并掌握服装产品系列策划和服装产品系列研发相关知识的术语与表达。

教学方式： 结合PPT与音频等多媒体教学课件，以教师课堂讲解为主，学生随堂练习与小组讨论为辅。

教学要求： 1. 掌握相关词汇。

2. 能够用英语描述出服装产品系列策划和研发的过程。

课前（后）准备： 课前预习关于服装产品系列策划和服装产品系列研发的基础理论知识，浏览课文；课后记忆单词，并复述文中内容。

12.1 Putting together a line
服装产品系列策划

The term "line" is used for moderate and popular-priced apparel, while the term "collection" is used to describe an expensive line in the United States or in Europe. A line is the garments conceived and designed, released and marketed for a particular season by an apparel company. It consists of an entire season's products including accessories as well as garments. A line can be divided into several groups of garments, linked by a common theme like color, fabric, or style.

"产品系列"（line）一词通常用于中档或大众化价位的服装，不同于"精品系列"（collection）在美国或欧洲用以形容高档的服装产品。服装产品系列是由服装公司在特定的季节构思设计、发布和销售的服装。它由当季的全部产品构成，既包括服装也包含配饰。一个产品系列可以分为几组服装，它们在色彩、面料或风格方面贯穿着同一个主题。

When the company and its designers begin to develop a line, they need a wealth of information to guide this creative process. They must identify what kind of clothes they want to design, for example, for men, women or children. They should also understand the different types of garment and whether these garments can form a line on their own or in combination. Finally they must consider strategically the different ways that they can promote and sell the line.

当服装公司和设计师开始研发服装产品系列的时候，他们需要大量的信息来引导这个创作过程。他们必须确定下要设计什么样的服装，例如，是为男士、女士还是儿童设计。还应了解不同类型的服装，以及这些服装是单独形成一个产品系列，还是组合而成一个系列。最后他们必须从战略上考虑产品系列推广和销售的不同渠道。

The fashion year has two seasons, six months apart. The industry works on a cycle, with one line for the Spring/Summer and another for the Autumn/Winter seasons. Small companies produce just these two lines a year, but larger companies produce more. Often, they sell two smaller lines for the Christmas period and high-summer. The Christmas line can include party wear or clothes for winter holidays. The high-summer line focuses on swimwear and summer holiday clothes. In order to drive sales, some large companies offer an almost continuous production of new styles. These fast-fashion models adhere to a much shorter time frame for development and production, in part enabled by technology.

时装年度一年有两季，以六个月为间隔。服装行业按周期开展工作，春夏季和秋冬季各发布一个服装产品系列。小的公司每年只生产这两个主系列，而大公司要生产得更多，通常他们还会在圣诞节和盛夏期间加推两个小型产品系列。圣诞系列可以包括派对服装或冬季度假服装。盛夏系列集中在泳装和夏季度假服装上。为了推动销售，一些大公司几乎连续不断地生产新款式，这些快时尚模式遵循更短的研发和生产时限，一定程度上还需要技术的支持。

Small pre-lines are produced as tasters of what is to come, and are shown to buyers just before the main lines. Designers may also produce a commercial selling line, from which buyers can primarily place their orders. This allows the main line catwalk shows to be more experimental and so catch the eye of the press.

小的季前系列生产出来作为当季产品的试卖款，先于主系列展示给经销商。设计师也许还会推出一个商业销售系列，在这里经销商基本可以完成他们的订单。这样使主系列产品在展示时具有更多的尝试性，便更能引起媒体的关注。

A designer may be working on many lines at one time. For example, in January a designer may be showing a pre-line, finishing the look of the Spring/Summer main line, finishing the high-summer line and starting to design the main Autumn/Winter line all at the same time. For a large read-to-wear company, the Spring/Summer line may have around 160 pieces, the high-summer line 100 pieces and the Autumn/Winter line 200 pieces.

设计师可能会在同一时间段内兼顾多个产品系列的工作。比如，一月份设计师可能在展示季前产品系列，完成春夏主系列的外观设计，完成盛夏系列，并开始设计秋冬主系列的服装，这些都是同时进行的。对于一个大的成衣公司来说，春夏系列大概有160件，盛夏系列有100件，而秋冬系列有200件服装。

12.2 Mood boards
概念展板

A useful way of assembling the many inspirational components is to make a mood board. Like an overture before a film begins, a mood board signals what is to come through ideas, themes, pictures, colors, and words. A mood board is used to introduce the design research and ideas to the team staff at the beginning of a line's development. It is a visualized presentation of mental work that the designer has done. It helps direct and explain design style, allowing others to imagine the effect that the designer is attempting to create; and is also convenient for designers to communicate with colleagues and clients.

将多个灵感组件汇合到一起的有效方法是制作概念展板。就像电影开始时的序幕一样，概念展板通过理念、主题、图片、色彩和文字来传达想要展示的内容。概念展板是在服装产品系列研发之初，用来向团队成员介绍设计调研和理念的。它是设计师所做的脑力工作的视觉呈现。它有助于引导和解说设计风格，让其他人想象设计师要创造的效果，还便于设计师与同事和客户之间进行沟通。

To generate the mood boards, designers often start with the resource files they have already compiled. They can collect tear sheets from magazines and printouts from websites, as well as swatches of colors and textiles, and to keep notebooks of written ideas and sketches (Figure 12-1). If space allows, a designer can place all the elements of the resource file—fabric swatches, sketches, photographs, buttons, trim,

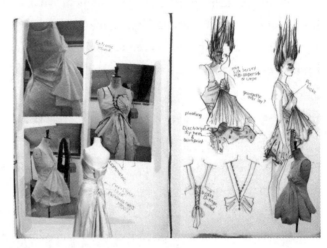

Figure 12-1　A designer's notebook
图12-1　设计师的笔记本

tear sheets, and printouts etc.—on a bulletin board or foam board (Figwre 12-2). In some studios, the assemblages can even take over an entire wall. Designers can step backward to see how these components correlate, looking for connections and contradictions among them; and extract something useful for the season's line. Finally, after collating these materials designers show the most distilled parts to others.

为了生成概念展板，设计师经常从他们已经收集的资源文件开始。他们可以收集杂志单页，打印网站上资料，还可以采集色卡和纺织品样本，并保存好记录着想法和草图的笔记本（图12-1）。如果空间允许，设计师可以把所有的资料——面料小样、设计草图、照片、纽扣、装饰配件、单页和打印成品等元素安排到公告板或泡沫板上（图12-2）。在有的工作室里，这些资料甚至会占据一面墙。设计师可以退后一些来观察

Figure 12-2　Mood board
图12-2　服装展板

这些元素之间是如何相互关联的，寻找它们的联系性和矛盾性；并提取出对当季产品系列有用的东西。最后，经过对这些材料的整理，设计师把最精华的部分展示给他人。

12.3 Developing a line
服装产品系列研发

Designers, often in conjunction with the merchandisers, review the research and past business reports to determine a theme, silhouette, and colorway along with a product assortment that will yield the desired profits. If the products of a previous season sold well, designers are likely to continue the style or develop on the basis of it instead of changing the style dramatically. So many companies will maintain a sense of evolution from the previous season when developing new products. While consumers are generally looking for something new, they are more likely to accept change when it occurs gradually and is related to something they are currently wearing. Additionally, designers should also consider a multitude of variables during the process of predicting the future, and unspoken desires of the consumer.

设计师经常与业务员一起回顾调研内容和以往的商业报告，来确定主题、廓型和色彩搭配，以及会产生预期利润的产品组合。如果上一季产品销售得好，设计师很可能会继续沿用这种风格，或在此基础上发展，而不是骤然改变这种风格。所以很多公司在研发新产品时会保持对上一季产品的延伸感。虽然消费者一贯在寻找新鲜事物，但当这种新鲜感悄然发生并与目前的穿着相关时，他们更易于接受这种变化。另外，设计师还应考虑到未来预测过程中的大量变数和消费者潜在的需求。

In large companies a line can have multiple groups totaling several hundred pieces. The type and number of an individual style are determined by the desired matching effect with other items. Designing parameters are set around cost, style, color, and type of the garment. For example, a simple line may consist of two pants, one skirt, one dress, two jackets, and three tops to be offered in two colorways, using two solid and one printed fabrication for each. With these parameters, the designer can then begin to design the specific style of each garment. Meanwhile the business manager guides designers by setting a focus and specific commercial goals.

在大公司的一个服装产品系列里会拥有多个产品组合，共计几百件服装。单个款式的类型和数量是由其与其他单品所期望达到的搭配效果决定的，其设计参数围绕服装的成本、风格、色彩和品种制定。例如，一个简单的产品系列可能由含有两种配色方案的两条裤子、一条裙子、一件连衣裙、两件夹克和三件上衣组成，每种配色方案运用两个纯色和一个花色的组合。有了这些参数，设计师就可以开始设计每件服装的具体款式了，同时业务经理通过设定重点和具体的商业目标来指导设计师。

Having digested the research and designing parameters, the line is designed and drawn up. Frequently, a designer works from designs that were strong sellers during a previous season. Modifying a style by changing the details, the color, or the fabrication can reduce the time and cost of developing garments for a new line. For example, last season's jacket may get a different fabrication, a change to welt pockets, and new buttons, and so on. In the meantime, designers should also think about selecting the fabrics as it contributes the most to the final cost and also has the most impact on the garment's look, performance and production.

在对调研内容和设计参数理解消化了以后，就开始设计和绘制服装产品系列。一般情况下，设计师从前一季的畅销款设计中开始工作，通过改变细节、颜色或制作来调整款式，可以节省研发新系列服装的时间和成本。比如，把上一季的夹克用不同的方法制作，改变一下贴边口袋，用新的纽扣进行装饰等。同时，设计师还应考虑面料的选择，因为面料是决定最终成本的最重要因素，也对服装的外观、性能和生产影响最大。

Next, designers draw a working sketch for each style, and select fabric samples and trims. For those important styles which can better express the designers' ideas and creativity, they will also draw fashion drawings. These works can be drawn by hand or using computers and are usually done by the assistant designers; the chief designer will control the entire designing direction and review all the design works. Assistant designers need to draw many more sketches than will actually be selected. The chief designer selects outstanding ones from these designs, so 100 sketches may be pared down to 50. Then these 50 works will be reviewed by the company team in the next step.

接下来，设计师对每个款式都要绘制服装款式图，并选择面料小样和设计辅料。对于那些主要的、能更好地表达设计师想法和创意的款式，还需绘制效果图。这些效果图可以手绘或运用电脑绘制，通常是由助理设计师完成的；首席设计师会把握整个设计方向，并审查所有的设计作品。助理设计师需要绘制比实际选用更多的草图。由首席设计师从中挑选优秀的设计，所以100个草图可能会削减到50个。然后这50个作品再由公司团队进行后面的评审。

All the proposed works will be presented on a designing board or in PPT. These designs are reviewed for their individual strengths, their relationship to the line concept, and their estimated cost. An open, objective critique is essential to pull together a strong line. A review meeting usually consists of designers, the boss, department managers, and technical designers; and some of the buyers may be also invited to attend the meeting. The review is focused on the designs, and the criteria set for their evaluation are taken from the original line concept and business plan. The critique is not about the designer, and personal attachment to the designer's work must be dismissed. The review process can take several iterations to identify the garments for each group within the line. Maybe the number of styles is insufficient or some pieces are missing, so further design ideation is necessary.

被推荐的作品全部都会在设计展板上或PPT中展示出来。相关人员会根据这些设计各自的优势，与整个产品系列理念的关系，以及成本评估来进行评审。公开、客观的评判对于组建一个有竞争力的服装产品系列来说非常重要。评审会一般是由设计师、老板、部门经理和工艺设计师组成；一些经销商可能也被邀请来参加这个会议。评审的重点放在设计上，其评估的标准从最初的产品系列理念和商业计划中制定。评判不是针对设计师的，而且对设计师作品的个人依恋必须要摒弃。评审的过程会经过几个回合，以甄选出产品系列里面每一组的服装。评审后，也许款式的数量不够或发现有遗漏的款式，因此还有必要进行二次设计构思。

Revisions due to cost are frequent. Fabric and labor costs must be controlled to achieve the desired price point. For example, a full skirt may require less expensive fabric or a reduction in the fabric amount that is needed to meet the established sale price. If a solution cannot be found, the garment is usually dropped from the line. The final selection of garments is made into samples for further review.

因成本问题而修订款式会很频繁。面料和人工成本必须要得到控制，以达到预期的价格水平。比如，一条蓬蓬裙会使用比较低廉的面料，或减少面料总量以符合既定的价格要求。如果找不到解决方案，这件衣服通常会从产品系列中撤下来。最终选定的服装新款会做成样衣进行下一步的评审。

Before a pattern and sample can be developed, the garment's techniques and trim details must be decided. This is done by the designer, or in large companies it becomes the responsibility of a technical designer. The details of the garment must be clearly specified. These include the final fabric, understructure, closure types, trims and accessories, edge finishes, seam and stitch types, labeling, hangtags, and packaging. Size specifications also need to be established in advance, while garment measurements can be determined after garments are finally approved. After samples have been made, it may be altered in terms of fit, fabrication and detail, and then re-sampled. The entire sampling process may take place in-house or it may be sent out to the factory that manufactures sample clothes. The designer or other members of the team will be in charge of this process.

在板型和样衣做出来以前，服装的工艺和装饰细节必须要确定下来。这项工作由设计师完成，或者在大公司里就变成了工艺设计师的责任。服装的细节要明确说明。这包括最终的面料、内部结构、闭合方式、装饰配件和服装辅料、边缘处理、缝合和缝纫的类型、标签、吊牌和包装。号型规格也要提前建立，服装尺寸可以在服装最终获得通过以后制定。样衣制作完成以后，可能会根据其适身程度、制作和细节进行调整，然后重新打样。整个样衣的制作过程可能是在公司内部进行，也可能送到样衣的制作厂家。设计师或其他的团队成员将负责掌控这个过程。

本章小结

■ 时装产品系列一年分两季，有的公司还会增加季前系列、圣诞系列和盛夏系列等。

■ 概念展板是展示设计研究内容和理念的很好方式，还便于设计师与同事和客户进行沟通。

■ 研发服装产品系列要经过调查研究、设计和绘制、评审、修订和确定工艺细节等过程。

Exercises 课后练习

1. Please discuss with your partner how to put together a line

请与你的同伴讨论如何策划一个服装产品系列。

2. Please talk about what is the use of a mood board.

请谈一谈概念展板的用途是什么。

3. Please discuss with your partner how to develop a line.

请与你的同伴讨论如何研发一个服装产品系列。

Chapter 13
第十三章

The Process of Garment Producing
服装生产过程

课题名称： The Process of Garment Producing

课题内容： 服装生产过程

课题时间： 2课时

教学目的： 让学生学习并掌握关于服装平面制板、立体裁剪，样衣制作与评估，服装生产等内容的叙述和表达。

教学方式： 结合PPT与音频等多媒体教学课件，以教师课堂讲解为主，学生随堂练习与小组讨论为辅。

教学要求： 1. 掌握相关词汇。

2. 能够用英语描述出关于服装制作、评估和生产等相关内容。

课前（后）准备： 课前预习关于服装制作、评估和生产等内容的基础知识，浏览课文；课后记忆单词，并复述文中内容。

13.1 Flat-patternmaking and draping
平面制板和立体裁剪

Construction is the foundation of clothing and fashion design. It is important for every fashion designer to understand how garments are made; this is both a technical and a design issue. Garments need seams and darts in order to render a two-dimensional fabric into a three-dimensional piece of clothing, but where and how a designer chooses to construct these lines also affects the proportion and style of a garment. To solve the issues of a garment construction, there are two ways: flat-patternmaking and draping.

结构是服装和服装设计的基础。对每个设计师来说了解服装是如何制作的非常重要，这既是一个技术问题，也是一个设计的问题。服装需要有接缝和省道，使二维的面料变成三维的服装，但是设计师选择在哪里以及如何设置这些线条也会影响到服装的比例和款式。解决服装的结构问题的方式有两种：平面制板和立体裁剪。

13.1.1 Flat−patternmaking 平面制板

Before the garment is made out of fabric, a pattern for the garment is cut out of paper, and this paper pattern is then used to cut the cloth for the garment. This work is flat-patternmaking. The pattern is drafted using a set of measurements, including circumference measurements and length measurements. The relationship of lines drafted on paper according to these measurements should also be considered: how, where, and at what angles they intersect. Patterns can also be taken from fabric pieces that have been draped on a dress form in order to develop a design. The more detailed information that is collected the better the fit.

在用面料制作服装以前，先用纸裁出服装样板，然后再按这个纸样裁剪服装布料。这项工作就是平面制板。板型用一组尺寸绘制：包括围度尺寸和长度尺寸。还应考虑根据这些尺寸在纸上绘制的线条的关系：它们以什么方式，在什么位置，以什么角度相交。为了达到设计效果，板型也可以从在人台上立裁的布片中提取。收集的信息越详细，服装就能做得越合体。

Next, ease is introduced to areas that require extra flexibility, e.g. across the back, at the armhole, and across the seat; then darts are positioned as fold lines where needed to eliminate fullness, and finally seam allowance is added beyond the stitching lines. Thirdly, a piece of muslin is pinned or basted together for a fitting to make corrections and refine the final pattern.

其次，在需要额外活动量的区域增加松量，如后背、袖窿和坐围处；然后在需要去除余量的地方设置折叠线条的省道，最后在所有缝线以外增加缝份。再次，把白坯布用珠针别成或假缝成试样衣，以修正和完善最后的板型。

13.1.2　Draping 立体裁剪

Some clothes are too complicated or innovative to be designed in two dimensions; these ideas need to be worked out physically in three dimensions by manipulating and draping fabric on a dress form (Figure 13-1). A creative designer can drape a cloth into a dress on a dress form merely by a pair of shears and a few pins. This process is just draping. Some designers prefer to work in this manner, because it allows the designers to really experiment with various forms. The possibilities of draping a garment on a dress form are arguably endless, limited only by the imagination.

有些服装太复杂或太有创意而不能用二维的方式设计，这些想法需要通过在人台（图 13-1）上操作和披挂布料，在三维的空间里以实物的形式实现。一个有创造力的设计师仅仅用一把剪刀和几个珠针就可以把布料在人台上做成一件衣服。这个过程就是立体裁剪。有些设计师喜欢用这种方式工作，因为它可以让设计师们真正尝试到各种造型。在人台上能够设计出的服装可以说是无穷无尽的，受限的只有想象力。

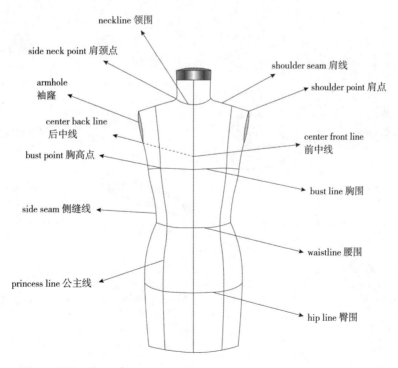

Figure 13-1　Dress form
图 13-1　人台

Manipulating draping on a dress form can help designers visually see the form and design effect of a garment; it can also provide them with a better sense of the body and bulk of a fabric, affecting decisions about finishing details. Understanding fabric tactile sensation and properties is essential to the success of a design; so draping is a good way to feel how the fabric breaks, folds, and drapes. Toolkit for flat-patternmaking and draping is shown as figure 13-2.

在人台上操作立体裁剪可以让设计师直观地看到服装形态和设计效果；还可以让他们对面料的体态和团块有更好的感觉，这会影响到对细节的处理。了解面料的质感和性能对设计的成功至关重要，立裁是一种很好的感受面料如何断裂、折叠和悬垂的方式。平面制板和立裁的工具箱如图13-2所示。

Further reading: Darts

Darts create fit. They are triangular, tapered or diamond shapes that, once folded out of a paper pattern or fabric, convert a two-dimensional shape into a three-dimensional form. Imagine a circle; by cutting out a triangle from it and then folding it, the two-dimensional circle becomes a three-dimensional cone. By suppressing fabric, the dart helps to shape fabric to the form of the body. Darts commonly point towards the bust (or bust point) and from the waist towards either the hip or bottom.

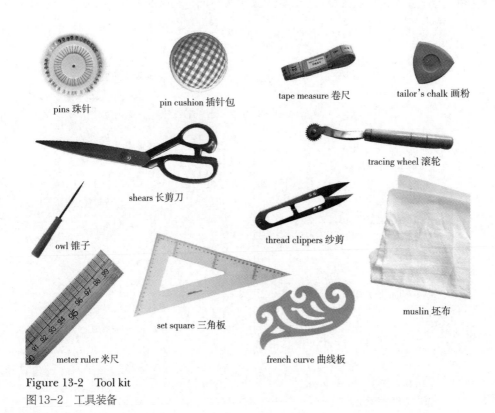

pins 珠针

pin cushion 插针包

tape measure 卷尺

tailor's chalk 画粉

tracing wheel 滚轮

shears 长剪刀

thread clippers 纱剪

owl 锥子

set square 三角板

muslin 坯布

meter ruler 米尺

french curve 曲线板

Figure 13-2　Tool kit
图13-2　工具装备

延伸阅读：省道

省道创建合体性。它是三角形，锥形或菱形的，一旦从纸板上或面料上折叠起来，它就会把一个二维的形状变换成一个三维的造型。想象一个圆形，从中剪去一个三角形然后折叠，二维的圆就变成了三维的锥体。通过折叠面料，省道有助于将面料塑造成人体的形态。省道通常指向胸部（或胸高点），或从腰部指向臀部或底摆。

13.2 Sample development
样衣研发

The first version of a garment made in final fabrics is called the "sample"; it is the garment that goes on the catwalk or is shown to the press. Many companies make their own samples, with designers working in close contact with the sample makers. But some companies send out the designs and develop specifications for a supplier to complete the sample-making process.

用最终选定的面料做出的第一个服装样本叫"样衣"，也就是在T台上展示或呈现给媒体的服装。许多公司都是自己制作样衣，设计师与样衣师紧密配合工作。但也有的公司将设计和制作规格外包给生产供应商来完成样衣制作。

Once the sketch has been approved, a patternmaker develops a pattern. The garment is cut and then sewn by a sample maker. The sample maker needs a wide range of knowledge and skills as she will complete construction of the entire garment. Her knowledge and skill are invaluable to the patternmaker and the designer as she can identify problems and offer solutions as the garment is developed.

设计图一旦获得了认可，制板师就开始打板，样衣师裁剪、缝制样衣。样衣师需要具备广泛的知识和技能，因为她要完成整件衣服的构造。她的知识和技能对制板师和设计师来说非常重要，因为在服装研发过程中她可以找出问题所在并提出解决方案。

Once the designer receives a first sample, he or she must measure and record the sample thoroughly, thus the designer can know the exact measurements of the existing garment before making changes. When a second sample is in hand, it must be thoroughly measured again to see if it fits the requested specs (specifications). Many houses plan to have three rounds of sample fitting in the season's development to ensure all garments are being produced as required. Garments that cannot be satisfactorily developed within the time frame allowed will be dropped from the line.

设计师一旦收到第一次的样衣，就要对样衣进行详细的测量和记录，这样设计师就可以在做出修改以前知道现有服装的准确尺寸。当收到第二次的样衣时，必须再次对其进行详细测量，看它是否符合要求的规格。很多公司都是在一个季度的研发中计划进行三次试衣，以确保所有的服装都是按要求生产。如果服装不能在允许的时间范围内被满意地研发出来，将会从产品系列中撤掉。

Retailers can void a contract if garments they have agreed to purchase are not meeting the specs that were requested. A spec sheet will indicate how much tolerance for error is allowed for each measurement. Specs exert a lot of control over fit. An approved sample and spec will be graded up and down from the sample size into the full range of sizes.

如果零售商订购的服装没有达到要求的规格，他们可以解除合同。规格表中会表明每一个测量尺寸所允许的公差范围。规格确保服装的合体程度在严格的控制之中，通过验收的样衣和其尺寸规格将从样衣尺码往大和往小进行推板，以涵盖整个型号范围。

13.3 Evaluation and adjustment
评估和调整

Once the samples are completed, a fit session is held to evaluate the design and the fit of the garment on a fit model. Samples prepared by suppliers are measured against the developing specifications. These are the exact dimensions of the garment at specific locations such as the waist or sleeve length. The garment must be within a specified tolerance, for example, 0.6cm. The fit session is attended by everyone involved in the development and production of the line including those in design, merchandising, patternmaking, technical design, production, and costing.

样衣制作完成以后，要召开试衣研讨会，来评估试衣模特身上所穿服装的设计和合体程度。生产供应商准备的样衣将按照研发规格进行测量。这些规格是服装上特定部位的精确尺寸，比如腰围或袖长。服装尺寸必须在规定的公差范围以内，如0.6厘米。每一个参与产品系列研发和生产的人员都要出席试衣研讨会，包括设计、销售、制板、工艺设计、生产和成本核算人员。

The garment is evaluated and approved, or suggestions are made to improve the garment's fit, appearance, or production. The model is asked to move around and give a personal report on the fit and comfort of the garment. Everyone is able to ask questions or point out potential problems that could affect the production and successful sale of the garment. Few garments are approved on the first review, though this is the goal because costs are greatly reduced. Most samples must go through more than one revision until approval is reached. Clear communication is critical so that revisions are easily understood by the person completing the pattern or production corrections whether they are in the next room or in another country.

研讨会成员对服装进行评估和通过表决，抑或提出改善服装合体性、外观或生产的建议。模特被要求走来走去，并给出关于服装合体性和舒适度的个人反馈。每个成员都可以提问或指出影响服装生产和成功销售的潜在问题。很少有服装能在初审时就获得通过，虽然这是大家的目标，因为成本会大大降低。在获得通过以前，大多数样衣须经过不止一次的修正。清晰无误的沟通是非常重要的，以便于完善样板或生产调整的人易于理解修正方案，无论他们是在隔壁，还是在另一个国家。

Technology is being developed that has the potential to reduce the development time for a sample. Fashion drawings can be rendered in 3D to allow better evaluation of the look of the garment in terms of proportions, silhouette, details, and colors. 3D body scanners are helping researchers develop sizing systems to improve fit for a wider range of body types. CAD patterns can be digitally placed onto a 3D digital avatar, giving information about fit without ever needing a garment to be cut and sewn. Fabric type and weight can be programmed to give a representation of the drape.

日益发展的技术有望减少样衣的研发时间。效果图可以用3D的效果渲染出来，便能更好地根据服装的比例、廓型、细节和色彩来评估它的外观。3D人体扫描仪正帮助研究人员开发尺寸数据系统，以为更多的体型改善合体程度。CAD板型以数字化的形式放置在一个3D的虚拟替身上，无须裁剪和制作服装，就可以给出关于合体性的信息。面料的类型和重量也可以通过编程的方式来模拟垂感的呈现。

13.4 Preproduction and mass production 试生产和批量生产

13.4.1　Preproduction 试生产

Once the final line has been approved, several steps remain before a garment is ready for production. If the line is to be sold wholesale, sales samples must be made and distributed to the sales representatives. Final production depends on the numbers of orders that are taken. The merchandisers and buyers from companies whose product goes directly to their retail outlets must finalize the number of sales to create an accurate order.

Garments can be dropped from the line owing to low orders.

最终的服装产品系列一经通过，在服装准备生产以前还有很多工作要做。如果服装产品是批发销售的，销售样衣须制作和分发给销售代表。最终的产量取决于获得的订单数量。如果公司的产品直接进入零售网点，其业务员和采购员必须最终确定销售的数量，以给出一个准确的订单。订单量少的服装会从产品系列中撤下。

Production patterns are made for these garments with profit-generating orders. A production pattern is engineered to exact size and style specifications and modified according to the final fabrication and equipment that will be used. For example, a pattern may need to be adjusted for expected shrinkage or for a special seam type. Markers are then made for each fabric type that will be cut, and these are designed to yield good fabric utilization. Markers provide the exact layout of the pattern pieces that need to be cut. Parameters can be set to control placement according to grain, nap, or match. Some companies use CAD to develop their patterns, and most use CAD to grade patterns and develop markers. The markers are then cut by hand or with an automatic cutter that is guided by electronic data from the marker. Preproduction and mass production activities are frequently sourced through a supplier or contract manufacturer if there are no company-owned production facilities.

订单中会产生利润的服装才会制作生产样板。生产样板被设计成精确的号型和款式规格，并根据最终采用的制作方式和生产设备进行修改。比如，样板可能需要根据预期的缩水量或特定的缝纫类型进行调整。然后为每个将要裁剪的面料类型制作排料，这些排料的设计是为了产生良好的面料利用率。排料可以给要裁剪的板型裁片提供精确的排列。可以根据面料的纹路、毛向或搭配设定参数，来控制排料的布局。一些公司用CAD研制板型，并且多数公司用CAD来推板和布置排料。然后是手动裁剪排料，或者用由排料的电子数据引导的自动裁切机裁剪。如果公司没有自己所属的生产设施，通常由生产供应商或合同制造商开展试生产和批量生产的活动。

In tandem with the preproduction activities, specifications for garments and all fabrics and materials must be finalized. Orders for fabrics and materials must be placed so that they arrive in time for production schedules to be met. Quality-control inspections take place at the production site, and garments are shipped directly to distribution centers or retailers.

在试生产活动进行的同时，服装的规格和所有的面辅料须最终确定。面料和辅料的订单必须提交，以使它们按时送达以满足生产进度的要求。在生产现场还要进行质量控制检验，之后合格的服装会直接运往分销中心或零售商处。

Each individual involved with the development, production, and marketing of an apparel line needs to be innovative and have a positive attitude. Ideas for improvements or new perspectives can come from anyone. While the designer may begin the process, those who produce and merchandise the line are invested in it as well. A small change may have an unanticipated chain reaction that affects the whole team both positively or negatively. The cutter who notices repeated flaws in the fabric before the cut begins can prevent an expensive error. The marker who notices that adding a center back seam can improve fabric utilization by 2% can reduce costs only if the designer approves the change. The sewer who continually experiences thread breakage on a new fabric needs help from an experienced production manager or the vendor who provided the fabric. If it was the designer who initiated the original fabric change, he or she needs to be aware of the impact it had on production and develop the foresight to avoid similar problems in the future.

　　每一个参与服装产品系列研发、生产和销售的人员都需要具有创新精神和积极的态度。任何一个人都可以提出改进的想法或新的观点。虽然这个过程可能是从设计师那里开始，但那些从事产品生产和销售的人也要投入其中。一个小小的改变可能会产生意想不到的连锁反应，进而对整个团队产生积极或消极的影响。裁剪师在裁剪开始前注意到面料反复出现瑕疵，就可以避免代价昂贵的失误；排料师注意到添加一个后背中缝可以提高2%的面料利用率，只要征得设计师的同意，就可以降低成本。缝纫师在使用新面料制作时连续断线，就需要得到经验丰富的生产经理或面料供应商的帮助。如果设计师想要改变原先的面料，他或她需要意识到这将对生产产生的影响，并且要有先见之明，以避免将来出现类似的问题。

13.4.2　Mass production 批量生产

When a garment is slated for production, it is graded to each of the various sizes in which it will be made. After a pattern has been graded into the various sizes, the pieces are laid out on a long piece of paper, which is the marker making. The marker is laid on top of the layers of fabric which are spread on a cutting table (Figure 13-3). Then the fabric

Figure 13-3　Cutting table
图13-3　裁床

will be cut into a garment's stitching pieces according to the marker, such as the sleeves, collars, fronts, and backs. Once the cutting is completed, the pieces of each pattern are tied into bundles according to their sizes. This work must be done by hand. Finally these bundles are sent to the manufacturer's sewing operators (Figure 13-4).

当一件服装准备投入生产时，它会被推板至将要制作的各种型号。当板型缩放出各种型号以后，板片被排放在长长的纸上，也就是制作排料。排料样片被铺放在裁床（图13-3）上布料层的上面。然后根据排料将面料裁成衣服的裁片，如袖子、领子、前片和后片。一旦裁剪工作完成，每个板型的裁片根据它们的型号被捆绑在一起。这项工作必须手工完成。最后这些捆绑包被送到生产商的缝纫人员手里（图13-4）。

Figure 13-4　Mass production
图13-4　批量生产

When making apparel during the production process, many factories use industrial sewing machines which work efficiently, and perform specialized functions. Some machines sew only seams, while some sew blind hems; button machines sew on buttons and overlockers finish raw edges (Figure 13-5). A completely automated sewing assembly line is under development in Japan. The only thing humans will do in this new system is supervise.

在制作服装的生产过程中，许多工厂都是使用生产效率高，并具有专业化功能的工业缝纫机。有的机器只缝接缝，有的机器缝制翻边；纽扣机缝纽扣，锁边机处理毛边（图13-5）。日本正在研制一条全自动的缝纫流水线。在这个新系统中，人们唯一需要做的就是监督整个过程。

industrial flat-bed machine 工业平缝机

steam ironing machine 蒸烫机

buttonhole machine 锁眼机

overlocker 锁边机

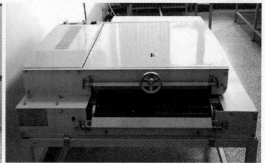

fusing press 热熔粘合机

Figure 13-5　Machinery
图13-5　机械设备

本章小结

- 平面打板和立体裁剪是解决服装结构问题的两种方式。
- 设计图获得认可以后，需要先研制样衣。
- 样衣完成以后，需要召开试衣研讨会，并对其进行评估。
- 那些会产生利润的服装将进入试生产，确定服装规格和所有材料，并进行统一的裁剪和缝纫。

Exercises 课后练习

1. Please interpret the features of flat-patternmaking and draping.

请阐述平面制板和立体裁剪的特点。

2. Please discuss with your partner how to develop and evaluate samples.

请与你的同伴讨论如何研发和评估样衣。

3. Please discuss with your partner about the process of preproduction and mass production.

请与你的同伴讨论试生产和批量生产的过程。

Fashion Designer
服装设计师

课题名称： Fashion Designer

课题内容： 设计师要具备的素质和能力

著名设计师分析

课题时间： 2课时

教学目的： 使学生了解成为设计师所需要具备的基本技能及熟悉设计师的风格

特征。

教学方式： 结合PPT与音频等多媒体教学课件，以教师课堂讲解为主，学生随

堂练习与小组讨论为辅。

教学要求： 1. 掌握相关词汇。

2. 能够用英语描述出设计师作品的风格特征。

课前（后）准备： 课前预习关于服装设计师的基础知识，浏览课文；课后记忆

单词，并复述文中内容。

14.1 Abilities required for a fashion designer
成为服装设计师所具备的能力

14.1.1　Solid foundation of skills 扎实的基本功

Fashion drawing and design skills are the basic professional qualities required for designers. In order to turn their ideas into reality, the designers need a combination of excellent drawing, cutting and dressmaking skills in addition to a wide knowledge of features of clothing fabrics.

服装绘画与造型能力是设计师必须具备的专业素质。要将脑海中的构思转化为实物，设计师除了要广泛了解服装的面料特点以外，还要有出色的绘画、裁剪和制作服装的综合技能。

14.1.2　Profound artistic attainments 深厚的艺术造诣

Due to the fact that fashion design is a process of artistic creation, artistic accomplishment and aesthetic senses are essential for fashion designers. The higher their artistic attainments, the richer their imagination and the greater their creativity. Many famous international fashion designers are outstanding artists.

服装设计是一种艺术创作，因此良好的艺术修养及审美对于服装设计师来说至关重要。深厚的艺术造诣决定了设计师的想象力和创造力。国际上许多著名的服装设计大师都是才华卓著的艺术家。

14.1.3　Divergent thinking 开阔的思维

Imagination and creativity are the soul of fashion design. As a reflection of their creative thinking, fashion works are the direct representation of designers' design ability. Only by broadening their mode of thinking can they be better inspired and create more excellent works.

想象力和创造力是服装设计的灵魂。服装作品体现了设计师的创意思维，直接反映出设计师的设计水准。只有拓宽思维模式，才能更好地激发灵感，创造出更多的优秀作品。

14.1.4　Unique viewpoints 独到的见解

Fashion design in fact is also a process of model selection, in which successful designers will not go with the flow, but develop their own opinions and understanding by continuously absorbing and learning from the latest fashion trends to form their own design styles.

服装设计实际上是进行式样选择的过程，在这个过程中，成功的设计师不会随波逐流，他们会不断地吸收和借鉴流行信息，形成独特的见解和主张，从而塑造独有的设计风格。

14.1.5　Keen market insights 敏锐的市场洞察力

The value of a fashion designer is ultimately to be tested in the market. Only the ones with keen insights can know the market and the needs of consumers timely to keep an invincible position in the competition.

服装设计师最终要在市场中检验其价值。只有具备敏锐的洞察力，才能及时地了解市场，了解消费者的需求，而使自己在竞争中立于不败之地。

14.1.6　Perfect commodity planning ability 完善的商品策划能力

Designers must have excellent planning ability. They have to draw up a theme before each season's new clothing design, and determine the style of their products according to popular elements of the year and thematic characteristics of their products. The entire planning process should be in the charge of the designers.

设计师要具备很强的策划能力。在每一季的新品设计之前，需要策划一个主题，根据当年的流行元素与主题特色，确定产品的风格。整个策划的过程都要由设计师进行把控。

14.2 Talent and hard work
天赋和汗水

With all the runway, flowers and applause, fashion designers in the center of the spotlights seem glamorous. But the efforts they have paid to become a high-profile designer are beyond common people's endurance. It is hard to be successful without distinguished talent, hard work and persistent enthusiasm.

T台、鲜花、掌声、镁光灯聚焦下的服装设计师风光无限。然而要成为一个众人瞩目的设计师，其光鲜背后的艰辛却非常人能及。如果没有傲人的天赋、勤劳的汗水和持久的热情，怕是难以成功。

Talent, which can endow the design with a soul, grows from the nurture of study and accumulation. Fashion design is a complex creation combining artistry and technicality. In order to complete a perfect work, designers have to look around for inspiration, be busy going to all kinds of fabric and accessory markets for materials, supervise the production process, examine the fitting, rehearse the show, hold press conferences and product promotions. In addition, the designers need to have in hand as much information as possible to improve their professional qualities, such as the abilities of keeping abreast of fashion trends, fabric and accessory information and the competitors' plans.

天赋可以让设计焕发灵性，但也需要后天的学习和积累。服装设计是一项艺术性与技术性相结合的复杂创作过程。为了完成一件完美的作品，设计师要四处寻找灵感，奔走于面辅料市场，督导服装制作，审视模特试穿，还要完成服装展示排演，召开新闻发布会以及产品推广会。除此之外，设计师还要尽可能多地掌握资讯，以提高自己的专业素质，如流行趋势、面辅料资讯，以及竞争者的动态等。

14.3 Celebrated fashion designers
著名服装设计师

14.3.1　French designers 法国设计师

Paul Poiret 保罗·波瓦列

In 1908 was the first groundbreaking dress unveiled by Paul Poiret who reformed women's clothing in a bold way by abandoning the use of corset which had been fettering women's bodies for centuries. He reproduced the natural beauty of human body by using soft silk fabric and oriental clothing style. His hobble skirt with a narrowed knee part was a great sensation at that time.

保罗·波瓦列在1908年推出了第一件突破性的女装，取消了几百年来束缚女性身体的紧身胸衣，大胆地对女装进行了改革。他选用柔软的丝绸面料，汲取东方的服饰风格，再现了人体的自然形态。他还推出了在膝盖附近收拢的"蹒跚裙"，曾轰动一时。

Coco Chanel 可可·夏奈尔

Coco Chanel (Figure 14-1) transformed the fashion world of the first half of the 20th century so completely that her designs could be found everywhere. As the creator of knit dressing, twin sets, little black dresses, Chanel had influenced the fashion world irrevocably. Black and white were her favorite colors for she believed that "simplicity" was the best way to make good textures of clothing. She was a pioneer in simple clothing styles and simple life styles as well, which made her the fashion leader of the era.

Figure 14-1　Coco Chanel
图14-1　可可·夏奈尔

可可·夏奈尔（图14-1）彻底改变了20世纪上半叶的着装时尚，她的设计无处不在。夏奈尔是针织女装、两件套、黑色小礼服等诸多经典款式的创造者，对时尚产生了持久有力的影响。黑与白是她最钟爱的色彩，她深信"简单"是让服装呈现美好质感的最佳方式。她倡导简约的服装风格和生活方式，成为引导时代潮流的一代大师。

Christian Dior 克里斯汀·迪奥

Dior (Figure 14-2) launched a "New Look" style in 1947. This style was characterized by round shoulder, full chest, tightened waist, and full A-line skirt, sweeping away the conservative silhouette of women's clothing after the World War Ⅱ, and winning great reputation for Christian Dior quickly. Dior's design focused on modeling lines, emphasizing the beauty of women's body rounded curves. Dior had been introducing various silhouettes from 1947 to 1957. His elegant yet varied apparel modeling and sculpture-like structure have great influence on fashion design even today.

Figure 14-2　Christian Dior
图14-2　克里斯汀·迪奥

　　迪奥（图14-2）在1947年推出了"新样式"

造型。这种造型以圆润的肩部、丰满的胸部、收紧的腰部以及膨起的A字裙摆为特点，一扫"二战"后女装保守呆板的线条，使他迅速成名。迪奥设计的重点在于塑造线条，强调女性玲珑有致的曲线美。在1947—1957年这十年中，迪奥不断推出各种服装廓型。他不断变化的优美服装造型以及雕塑般的服装结构，至今对服装设计都有影响。

Cristobal Balenciaga 克里斯特巴尔·巴伦夏加

Balenciaga created the simple and sporty outfit. He often hung or wrapped the fabric directly on models for designing and cutting, which got him the name of "Scissors Magician". From drawing, material selection, cutting and sewing to accessory selection, he was personally involved in every step of the process, and was a rare all-rounder in the fashion industry. In the 20th century, a Balenciaga gown cost much more than that of any other couturier. Balenciaga ignored trends or fads and his haute couture was never advertised, but he earned attention, respect, and deep loyalty.

巴伦夏加开辟了简约化的运动型服装。他经常直接将面料披挂或缠绕在模特身上进行设计和裁剪，因此被称为"剪刀魔术师"。从画图、选料、裁剪，缝制到选择配饰，每个过程他都亲力亲为，是时尚界少有的全才。在20世纪，巴伦夏加礼服比其他任何时装设计师的都要贵得多。巴伦夏加忽略了时尚和潮流，从未登过广告，却赢得了人们的关注、尊重和忠诚。

Yves Saint Laurent 伊夫·圣·洛朗

Yves Saint Laurent (Figure 14-3) was skilled in making up for figure deficiency, and integrating a variety of art and cultural elements into his design. Smoking dresses, see-through dresses and ethnic costumes of different countries were all his astonishing classic designs. As a leader in high-end ready-to-wear field, Yves Saint Lauren advocated the use of clothing to reflect the natural beauty of women's form. He said: "Fashion is not only to make women more beautifully, but to lead them spiritually and make them more confident."

伊夫·圣·洛朗（图14-3）善于修正人体体型的缺陷，并常将多种门类的艺术、文化元素融于他的设计中。吸烟装、透视装以及不同国度的民族风格服装都是他惊世骇俗的经典之

Figure 14-3　Yves Saint Laurent
图14-3　伊夫·圣·洛朗

作。他也是一位高级成衣的引领者，他主张通过服装来体现女性的自然形态。他说："时装并不仅是让女性更美丽，而是在精神上领导她们，让她们变得更加自信。"

Jean-Paul Gaultier 让·保罗·戈蒂埃

Gultier's enduring contribution to fashion is his inspection of gender roles and the integration of different religious cultures in design. He recommended women to wear bras outside to highlight the charm of women's chests. He is widely recognized as a bohemian prodigy in the fashion world. The clothing he designed incorporated the characteristics of humor, relaxation and wisdom. He insisted on dress equality between men and women, and strived to break through the clothing differences between men and women, which made his design quite neutral.

戈蒂埃对时尚持久的贡献，是他对性别角色的考察，以及将不同的宗教文化融于设计之中。他推崇将女性的内衣外穿，突出女性胸部的魅力。在时装界他被认为是一个放荡不羁的神童。他所设计的服装幽默、轻松、富有智慧。他坚持男女服饰平等的原则，努力打破男女服装上的差异，因此他的设计是相当中性化的。

14.3.2　Italian designers 意大利设计师

Valentino Garavani 瓦伦蒂诺·格拉瓦尼

In the 1970s and 1980s, Valentino (Figure 14-4) became the first couturier who simultaneously launched both men's and women's ready-to-wear garments, and creatively used letter combinations as decorative elements. He loved the purest colors, especially red, and bright red was the representative color of his design works. Oriental exoticism was often reflected in his designs. Valentino was always striving for perfection, so the elegant and beautiful shape, fine and comfortable fabrics as well as delicate and exquisite workmanship could be seen in every detail of his design.

在20世纪七八十年代，瓦伦蒂诺（图14-4）成为同时推出男式和女式成衣的首位高级时装设计师，并首创用字母组合作为装饰元素。他钟爱最纯粹的颜色，尤其是红色，鲜艳的红色可以说

Figure 14-4　Valentino Garavani
图14-4　瓦伦蒂诺·格拉瓦尼

是他设计的代表色彩。东方的异国情调也经常体现在他的设计中。瓦伦蒂诺力求尽善尽美，以高雅优美的造型、舒适精良的面料、精细考究的做工，贯穿于他服装的每一个细节中。

Giorgio Armani 乔治·阿玛尼

Although rarely related to "stylish", Giorgio Armani's clothing was elegant and reserved; it arose from his belief that clothing quality mattered more than styles. His best recognized master pieces of designs were the elegant suits for men and women, which were perfectly cut and well made, and made of exquisite fabrics. Therefore, as a symbol of high taste and quality, Armani's clothing was well-received among high-level intellectuals in the world. In his design there often appeared loose shoulder lines, soft silhouettes and subtle colors such as brown, purple, blue and cream. For women, Armani created a type of refined and soft suits, suggesting a quiet strength.

乔治·阿玛尼的服装风格典雅含蓄，却很少与"时髦"相关，因为他相信服装的质量更甚于款式。他最受推崇的代表作是优雅的男女西服套装，其裁剪合体，做工精良，面料考究。阿玛尼的服装是高品位、高质量的象征，因此深受世界高知人士的喜爱。在他的设计中经常出现宽松的肩线和柔和的轮廓，以及棕色、紫色、蓝色和奶油色等微妙色彩。对女性来说，阿玛尼创造的精致柔软的西服套装，蕴含着一种安静的力量。

Gianni Versace 詹尼·范思哲

Versace wanted women to feel sexy, powerful and independent by wearing his clothing. He represented it not only with dress shape but also through his distinctive design attitude—fashion had to be free to express individuality. Versace was known for making the best use of colors and for the skill of bias cut. His brand mark is the image of Medusa, a representation of fatal attraction in the Greek mythology. He had been so persistent in seeking the power of beauty in all his life that a sexy appeal was highlighted both in men's and women's clothing.

范思哲希望女性通过他的服装感受到性感、强大和独立。他不仅通过服装的造型来表达这一点，还有他独特的设计态度——时尚必须要自由地表达个性。范思哲以善于运用色彩和斜裁的技巧而闻名于世。他的品牌以蛇发女妖美杜莎为标志，美杜莎在希腊神话中代表着致命的吸引力。他一生都在追求这种美的震慑力，无论男装还是女装都突显出一种性感的味道。

14.3.3　American designers 美国设计师

Calvin Klein 卡尔文·克莱恩

Calvin Klein (Figure 14-5) strictly reserved artistic control over all his products, and "less is more" is his main design philosophy. Klein wanted to create wearable clothing for an increasingly female workforce. He focused on soft tailoring, neutral color palettes, and interchangeable pieces that helped women look professional but not stuffy. He used luxurious fabrics, such as cashmere, silk, and Italian wool, in silhouettes that were largely unadorned for an easy elegance. The formula worked well and became the hallmark look of any Calvin Klein collection.

Figure 14-5　Calvin Klein
图14-5　卡尔文·克莱恩

卡尔文·克莱恩（图14-5）严格地保留了他对所有产品的艺术话语权，并且"以少胜多"是他主要的设计原则。克莱恩希望为越来越多的职业女性设计实用的服装。他专注于柔和的裁剪、中性化的色彩和可混搭的单品上，使女性着装显得职业而不古板。他选用奢华的面料，如羊绒、丝绸和意大利的羊毛织物，服装廓型却不加修饰，以营造一种轻松优雅的感觉。这一造型效果很好，成为卡尔文·克莱恩所有高级服装系列的标志性形象。

Donna Karen 唐娜·卡伦

Donna Karen's design was rooted in the unique life style of New York and exuded the unique urban flavor of New York. The dresses she designed, from styles to colors, were full of mature female consciousness. For her, fashion design was a kind of self-expression. Her design was the simplification of life, simple but elegant, blended with feminine softness and energy, and full of metropolitan modernity.

唐娜·卡伦的设计植根于纽约特有的生活模式，散发着纽约特有的都市气息。其设计的女装从款式到色彩，都极富成熟女性的意识。对她而言，时装设计就是一种自我表达的方式。她的设计是对生活的凝练与简化，朴实无华却高贵典雅，既融合了女性化的柔美与活力，又充满了都市化的现代感。

14.3.4　British designers 英国设计师

John Galliano 约翰·加里亚诺

John Galliano (Figure 14-6) has enjoyed critical acclaim from the fashion press, and celebrity status among the fashionistas of the world, largely because of his ability to interpret the past in a larger-than-life way. He was entitled as the "Hopeless Romantic Master". His collections offer romance and theatrical imagination that affect all facets of the fashion industry. He once said that it was important for a designer to just be curious, and he or she mustn't be frightened or hide behind preconceived ideas. When he worked for the House of Dior, Galliano also focused on highly glamorous and exquisitely crafted gowns.

Figure 14-6　John Galliano
图14-6　约翰·加里亚诺

约翰·加里亚诺（图14-6）一直享有时尚界的赞誉和国际设计大师的巨星地位，主要是因为他有能力以超越生活的方式诠释过去。他被誉为"鬼才设计师"。他的作品充满了浪漫和戏剧般的想象力，其影响力渗透到了时装业的各个角落。他曾经说过，一个设计师拥有好奇心非常重要，而且一定不要害怕或躲避别人对自己的成见。在为迪奥工作期间，加里亚诺还曾专注于设计极具魅力、工艺精湛的礼服。

Alexander McQueen 亚历山大·麦昆

Alexander McQueen's impact on the fashion world is a force beyond trend or style dictate. Deeply influenced by street culture since childhood, he earned the title of "Hooligan of English Fashion" due to his unorthodox design. It seemed so easy for him to meld fashion, art, and concepts with extraordinary cutting skills and new technology. McQueen's works were full of drama and innovation. He was adept in taking ideas from the past and sabotaging them with his cut to make them thoroughly new. He subverted the aesthetic tradition and pushed the vision of fashion to a larger realm.

亚历山大·麦昆对时尚界的影响有一种超脱潮流或风格的力量。他自小受街头文化的影响，因其邪魅怪异的设计而赢得了"英国时装界坏小子"的称号。他似乎可以毫不费力地将时尚、艺术、概念，与超凡卓越的裁剪和新工艺融合在一起。麦昆的作品充满了戏剧性与创造性。他善于从过去汲取灵感，再大胆地加以"毁坏"，使它们焕然一新。他颠覆了传统意义上的审美，将时尚视野推入了一个更大的领域。

Vivienne Westwood 薇薇恩·韦斯特伍德

Deeply influenced by anarchism, Westwood (Figure14-7) was known for her deviance. She advocated clothes with new values and was popular with hippies; she was the pioneer of punk revolution as well. Her peculiar thought patterns were normally revealed in twisted stitches, asymmetric cuts, unfinished hemlines and uncoordinated colors. She successfully integrated street fashion into the field of fashion. She insisted that being sexy was fashion, so her designed dresses had always strongly stressed the chest and buttocks. Wearing bras outside and destructive design could be seen in her designs frequently.

Figure 14-7　Vivienne Westwood
图14-7　薇薇恩·韦斯特伍德

韦斯特伍德（图14-7）深受无政府主义的影响，以离经叛道而著称。她提倡新价值观的服装，深受嬉皮士们的追捧，她还是朋克革命的先驱者。其独特的思维模式通常表现为扭曲的缝线，不对称的剪裁，尚未完工的下摆和未调和的色彩。她成功地将街头流行融入时尚领域。韦斯特伍德坚持性感即时尚的观点，所以她设计的服装从来都是极力地强调胸部和臀部。内衣外穿和破坏性的设计随处可见。

14.3.5　Japanese designers 日本设计师

Issey Miyake 三宅一生

Issey Miyake (Figure 14-8) melded style, fashion, and wearability in a seemingly effortless way, which was what so many designers had hoped for but only a few could accomplish. Miyake had strived throughout his career to create a wearable fashion, but his artistry, ingenuity, and virtuosity had also been highly recognized. The clothes he designed were quite loose and of various styles. He was good at creating a variety of texture effects by exploiting the characteristics of fabrics to the maximum, which made his design always refreshing and earned him the name of "Fabric Magician".

Figure 14-8　Issey Miyake
图14-8　三宅一生

三宅一生（图14-8）能够做到许多设计师梦寐以求，却很少有人能做到的事情，就是将风格、时尚和可穿性以一种看似轻松随意的方式融合在一起。三宅一生一直都致力于创造可穿着的时尚，其艺术才能、独创性和精湛的技艺也得到了高度的认可。他的服装设计得很宽松，款式各异。他善于最大限度地利用面料的特点创造出各种肌理效果，这使他的设计总是能让人耳目一新，他也因此被称为"面料的魔术师"。

Yohji Yamamoto 山本耀司

Yohji Yamamoto (Figure 14-9) is a romanticist. Yamamoto's approach to creating alternative garments for women is based on combining Eastern and Western aesthetics. Along with his fellow Japanese designers, Yohji Yamamoto helped to redefine clothing and the use of color, shape, and form in relation to the figure. He has also helped to redirect and question the Western ideals of beauty and what it means to be a woman in today's society.

Figure 14-9   Yohji Yamamoto
图14-9   山本耀司

山本耀司（图14-9）是个浪漫主义者。他为女性创造非主流服装的方法，是在结合了东方和西方美学基础上的。和其他日本设计师一起，山本耀司重新定义了服装，并打破根据体型设计服装颜色、造型和形态的传统。他还帮助重新定位和质疑了西方的审美理想，以及女性在现代社会中的意义。

Rei Kawakubo 川久保玲

Designing under the label "Comme des Garçons" (meaning "like boys" in French), Kawakubo is the "Philosopher of Fashion" and one of the chief advocates of the deconstruction aesthetic introduced in the late 1970s. She creates clothing with a deeper commentary on image, the body, and sex appeal, and consistently challenges silhouette and presentation of fabric.

为自己的品牌"像男孩子一样"做设计的川久保玲是一位"时尚哲学家"，也是20世纪70年代末引入解构主义美学的主要倡导者之一。她创作的服装对形象、身体和性感有着更深刻的理解，并对服装的廓型和面料的外观呈现不断地进行挑战。

<h2 style="text-align:center">本章小结</h2>

- 成为一名合格的设计师要具备出色的专业技能和坚强的毅力。
- 了解国际著名服装设计师的设计风格。

<h2 style="text-align:center">课后练习</h2>

1. Collect materials after class and master the style characteristics of famous fashion designers.

课下搜集资料，掌握著名时装设计师的风格特点。

2. Group discussion 小组讨论

List your favorite designers. Explain the reasons and analyze their design style.

列举你喜欢的设计师，阐述其原因并对设计师的设计风格进行分析。

Bibliography
参考文献

［1］ SORGER R, UDALE J. The Fundamentals of Fashion Design [M]. second eition. Lausanne: AVA Publishing SA, 2012.

［2］ BYE E. Fashion Design [M]. Oxford: Berg Editorial offices, 2010.

［3］ TORTORA P G, EUDANK K. Survey of historic costume [M]. second eition. New York: Fairchild Publications, 2013.

［4］ CALDERIN J. The Fashion Design Reference & Specification Book [M]. Massachusetts: Rockport Publishiers, USA. 2009.

［5］ FAERM S. Fashion Design Course [M]. New York: Barron's Educational Series, Inc., 2010.

［6］ Hua Mei. Chinese Clothing [M]. New York: Cambridge University Press, 2011.

［7］ HALLETT C, JOHNSTON A. Fabric for Fashion: The Complete Guide [M]. London: Laurence King Publishing, Ltd, 2014.

［8］ SINGER R. Fabric Manipulation 150 Creation Sewing Techniques [M]. Newton Abbot: F&W Media International, Ltd., 2013.

［9］ HIDALGO M R, ROIG G M. Designing Fashion Accessories [M]. Pennsylvania: Schiffer Publishing, Ltd., 2012.

［10］ TATHAM C, SEAMAN J. Fashion Design Drawing Course [M]. New York: Barron's Educational Series, Inc., 2003.

［11］ VOLPINTESTA L. The Language of Fashion Design [M]. Massachusetts: Rockport Publishers, 2014.

［12］ MCKELVEY K, MUNSLOW J. Fashion Design: process, innovation & practice [M]. second edition. West sussex: John Wiley & Sons, Ltd., 2012.

［13］ JENKINS N. Style Is Eternal [M]. Victoria: Melbourne University Publishing, Ltd., 2014.

［14］ LOVINSKI N P. The World's Most Influential Fashion Designers [M]. New York: Barron's Educational Series, Inc., 2010.

［15］ 张玲. 图解服装概论 [M]. 北京:中国纺织出版社,2005.

［16］ 张蕾. 服装设计英语 [M]. 北京:化学工业出版社,2014.

［17］ 辛芳芳. 服装专业英语 [M]. 上海:东华大学出版社,2011.

［18］ 邓跃青. 现代服装设计与实践 [M]. 北京:清华大学出版社,2010.

［19］ 沈周. 古代服饰:英汉对照 [M]. 合肥:黄山书社,2012.

［20］ 田琼. 童装设计 [M]. 北京:中国纺织出版社,2015.

［21］ 周叔安. 汉英英汉服装分类词汇 [M]. 北京:中国纺织出版社,2012.

［22］ 李当岐. 西洋服装史 [M]. 北京:高等教育出版社,2005.

Picture reference
图片引用

图 1–2: https://www.dailyfashion.cn/lookbook.php?id=18646

图 1–3: http://www.sohu.com/a/211715011_99942937

图 1–4: https://huaban.com/pins/1115648465/

图 1–5: http://www.sohu.com/a/203911694_500120

图 1–6: https://fashion.qq.com/a/20160329/039207.htm

图 4–3: http://pic.haibao.com/image/15324301.html?kw=%E7%B2%89%E7%BA%A2%E8%89%B2&skip=84

图 4–4: https://huaban.com/pins/1058377900/

图 4–5: https://www.dailyfashion.cn/lookbook/34029

图 4–6: http://m.sohu.com/a/122524535_559321

图 5–8: https://lohas.pclady.com.cn/189/1894098.html

图 5–11: http://www.sohu.com/a/74207438_365546

图 6–1: http://www.fashiontrenddigest.com/d/5682.shtml

图 6–2: http://www.chinasspp.com/ 时尚品牌网

图 8–5: http://www.eeff.net/forum.php?mod=viewthread&tid=1137233&extra=page%3D1

图 9–3: http://www.sczdfs.com/show_product.asp?owen1=8&owen2=32

图 9–5: http://news.ggo.net/sscl/35_185350_1.html

图 12–1: 大众点评网

图 14–1: http://www.uglytruthofv.comwp–contentuploads201211coco–chanel

图 14–5: http://www.mydesignweek.euwp–contentuploads201412TOP–Fashion–Designers–of–all–time–calvin–klein

图 14–7: https://beyondmags.comwp–contentuploads201806interview–1

图 14–9: https://www.toodaylab.com76243